FEEDING ON DREAMS

BOOKS BY ARIEL DORFMAN

FICTION

Blake's Therapy

The Burning City (with Joaquín Dorfman)

The Nanny and the Iceberg

Konfidenz

The Rabbits' Rebellion

Mascara

The Last Song of Manuel Sendero

My House Is On Fire

Widows

Hard Rain

NONFICTION

Feeding on Dreams: Confessions of an Unrepentant Exile

Other Septembers, Many Americas: Selected Provocations

Desert Memories: Journeys Through the Chilean North

Exorcising Terror

Heading South, Looking North: A Bilingual Journey

The Empire's Old Clothes

*Some Write to the Future: Essays on Contemporary
Latin American Fiction*

How to Read Donald Duck (with Armand Mattelart)

PLAYS AND POETRY

Purgatorio

Picasso's Closet

Speak Truth to Power: Voices from Beyond the Dark

Reader

Death and the Maiden

In Case of Fire in a Foreign Land

FEEDING ON DREAMS

Confessions of an Unrepentant Exile

ARIEL DORFMAN

MARINER BOOKS

HOUGHTON MIFFLIN HARCOURT

BOSTON / NEW YORK

First Mariner Books edition 2012

For information about permission to reproduce selections from this book,
write to Permissions, Houghton Mifflin Harcourt Publishing Company,
215 Park Avenue South, New York, New York 10003.

www.hmhbooks.com

Library of Congress Cataloging-in-Publication Data
Dorfman, Ariel.
Feeding on dreams : confessions of an unrepentant exile / Ariel Dorfman.
p. cm.
ISBN 978-0-547-54946-0   ISBN 978-0-547-84418-3 (pbk.)
1. Dorfman, Ariel. 2. Authors, Chilean — 20th century — Biography.
3. Authors, Exiled — United States — Biography. I. Title.
PQ8098.14.O7Z467 2011
863 — dc22
[B]   2010049803

Book design by Melissa Lotfy

Printed in the United States of America
DOC 10 9 8 7 6 5 4 3 2 1

This book is for Angélica.

I know how men in exile feed on dreams.

— Aeschylus

A NOTE ABOUT TIME AND EXILE

PERHAPS IT was inevitable. Because I have lost my country three times in the course of one lifetime, the attempts at self-scrutiny that habitually accompany human existence have, in my case, been forced to grow and ripen through the fragmentation of many arrivals, returns, and departures, complicating the natural intricacies that every exercise in remembering, every memoir, already faces. Life may unfold chronologically for the body and for bureaucracies that keep track of such things as births, marriages, deaths, visas, tax returns, expulsions, and identity cards, but memory does not play this game in quite the same way, always manages to confound the desire for tidiness. And so, for readers who might feel perplexed as they accompany the protagonist in his struggle to find a way in and out of a labyrinth of recollections, a timeline, fixing the major events into a semblance of order, has been added at the end of this book. I don't necessarily recommend that readers consult that chronology. But how could I, of all people, deny to those who wander in the desert a stable glimpse of stars that can perhaps guide them to a safe haven?

INTRODUCTION

As a child, I would often dream of surviving my own death.

Lying there alone in the dark, I'd imagine what it would be like to watch my body stretched out on a bed, and everyone so sad, and all the while I'd be invisibly nearby, eager to jump out from the other side of immortality. Gone from this world for just a few hours and then mischievously alive again, ready to witness the bewilderment of the living when I'd resurrect, *Hey, look at me.*

Of course, when the opportunity finally did arise, when the day came many decades later and I heard a voice tell me that I was dead, that according to a newswire my body had been discovered that day, September 12, 1986, in a ditch on the outskirts of Santiago, hands tied behind my back and throat slit, it turned out that to witness my own death was less amusing than my childhood fantasies had anticipated.

Not that the news itself was that surprising. After all, I had been returning on and off to hazardous, dictatorial Chile for three years, ever since the military had allowed me back in 1983 after a decade in exile; so anything and everything could have happened to me there, but not now, not now that I was teaching a semester at Duke University in Durham, North Carolina, safely ensconced in the office where the reporter from United Press International had tracked me down — could I comment, please, on the fact that I had died?

"The reports of my death," I said to him, "have been greatly exaggerated."

A moment after I'd delivered that deliciously absurd sentence of Mark Twain's, I began to feel sick. The humor had put a cautious, witty distance between myself and somebody else's death, and postponed the need to ask the question: If I hadn't been murdered — at least not yet — then whose throat *had* been slit in that ditch in Santiago?

I learned the answer soon enough, but first had to respond to another telephone call — from my mother in Buenos Aires, desperate because she'd been contacted by some callous journalist seeking a statement on her son's assassination. Having reassured her that I was very much alive, my thoughts went out to another mother; there had to be a woman in Chile who could not explain away that killing, a woman who was at that very moment overwhelmed with grief, a woman, a wife, a sister. I had been infected enough by recent history to know that in times of tyranny it is mostly the men who do the dying and the women who do the mourning.

Somebody real had died in Santiago.

His name was Abraham Muskatblit.

On September 7, 1986, just five days before that macabre telephone call, a small ultra-left guerrilla commando had come within inches of slaying Chile's dictator, General Augusto Pinochet. A massive and peaceful movement in opposition to Pinochet had been growing in strength, and the military took advantage of that attack to clamp down on dissidents of all stripes.

As I watched the repression unfold from afar, my weird hope was that my friends in Chile would be taken to prison, the only safe place in a country where men in ski masks, bent on revenge, were breaking down doors and kidnapping citizens. By the afternoon of September 8, bodies began to turn up all over Santiago.

Among the dead was Pepe Carrasco. I had taught him the delights of *Don Quixote* at the University of Chile when he had been one of my students in the early sixties. Though he had afterwards abandoned literature for the more urgent career of journalism, we had not lost contact, nor our shared love for Cervantes. In fact, when we'd managed to meet the few times I'd been able to visit

Chile since 1983, I would jokingly call him Sansón, an allusion to another Carrasco, the barber in the *Quixote*.

"Sansón, Sansón," I had called out one afternoon in Santiago a few months before he was executed, when I'd noticed him at the edge of a crowd that had gathered for the funeral of a student killed by the police. Pepe was accompanied by two sons just arrived from Mexico, where the family had been exiled. "This is the man," he said to them, pointing at me, "who taught me to tilt at windmills." The extent of our last conversation. Tear gas and batons broke up the throng of mourners, but I'd managed to send a farewell gift to Pepe Carrasco, the same words for everybody in Chile those days: *Cuídate, Sansón*. I asked my friend to take care of himself.

As if by pronouncing those syllables, naming the danger, we could conjure it away.

Now it was his funeral being carried out in Santiago and I wasn't there to attend it. That the circumstances were so familiar only made them more painful. We seemed to be trapped in an incessant repetition of the 1973 coup, when the democratically elected government of Salvador Allende had been brought down. Now, again, thousands were being hunted down, and again it was impossible to ferret out their fate, even to speak openly on the telephone. The wife of one of my friends was enigmatic when I called from abroad. "Jaime let me know that he's sleeping with his clothes on," she said. I completed her words with my own thought, just in case . . .

Because I'd already been told that Pepe Carrasco's captors hadn't let him dress before they hauled him away. *No te pongas los zapatos*, they said to him. *No te van a hacer falta*. Don't bother to put on your shoes. You won't be needing them.

And two days later, the news of my fraudulent demise had arrived, flowing and bleeding into the slow and endless death of Chile itself.

I felt like a ghost, and not only because I could not intervene in what was happening in that country of mine to the south of no-where, change any of it. A ghost because I couldn't remember a time when there were other forms of passing away, when people died of old age or of sickness or in car accidents. Because everyone who had received the announcement of my execution hadn't for a moment doubted its veracity; every last person had considered a murder of this kind to be a most natural, indeed almost normal, occurrence.

Death had violently entered my life forever on September 11, 1973, the day of the military takeover. I had been able to escape the carnage due to a chain of miraculous coincidences that had kept me away from La Moneda, the presidential palace, where I'd been working as the cultural and press adviser to Allende's chief of staff. But luck would take me only so far. The Chilean Resis-tance—hammered out of the remnants of the Unidad Popular par-ties that had accompanied Allende's presidency and were fighting to continue to struggle under extreme duress—had ordered me to leave the country, and I had eventually, and reluctantly, made my way into exile. Nevertheless, I could never shake the sense that I was living on borrowed time, that a death was awaiting me in San-tiago. Barely a few months before I was informed of my execution in Santiago, Jesse Helms, an ultra-right-wing Republican senator from North Carolina, had denounced me on the Senate floor and offered a dossier on my travels that could only have been culled by Pinochet's secret police, could only have existed if those men were spying on me. Maybe I was in danger. Hadn't Pinochet targeted opponents in Mexico, blown up Orlando Letelier, Allende's minis-ter of defense, in Washington? Hadn't his neofascist allies stabbed a prominent Chilean in the streets of Rome? Was this a premo-nition of things to come, a night in my future when I would be transformed into a real ghost?

I fought back against my phantom circumstance with the weapon I had been brandishing since childhood to defeat the threat of my own extinction: I began to tell the story, transmit the lives of those who could not speak, either because they were dead like Pepe or because they were silenced like my imprisoned *compañeros.*

And I did so in English, the lingua franca of our era, which might give me access to the decision-making elite of the United States. And what of Spanish? The Spanish in which Pepe had died, the Spanish of his tormentors as they approached him, the Spanish that awaited me, along with my possible death, as soon as I returned to the Santiago of death squads, what of the Spanish spilling out of me and inside me? The Spanish that needed to tell the story as well, in the language of the victims, in the language of the perpetrators, so the community and the country would not forget, so the terror could be tamed with words? The Spanish of Borges suggesting that the future was an indecipherable book that we are unable to read until fate has brought us face to face with the man who will murder us? The Spanish of García Márquez and his hope that death might someday not be foretold, *una muerte anunciada?* The Spanish of García Lorca as he looked straight at the firing squad, the Spanish he whispered to name the ditch, *la zanja,* into which his body would be cast?

Spanish would get its turn. I promised my Spanish that I would get to its ambivalent sweet syllables as soon as I dispatched the article on my false death in English. The Spanish I had been born into had, by then, learned patience and tolerance, the advantages of sharing its breath with a rival tongue, had learned to cohabit in a civil way with the English zone of myself.

It had not always been like this.

It was in a New York hospital in the winter of 1945 and on a day I do not remember that I had renounced the language of my parents. An Argentine child of two and a half only recently arrived at the cold desolation of wartime Manhattan, I had caught pneumonia and, after three weeks of confinement in a hospital ward, had toddled out sane in body and probably insane in mind, unwilling to speak a word of my mother tongue, unable even to this day to recall the loneliness I must have endured under those white monolingual walls where I embraced English forever. Though *forever* is a dangerous word for anyone belonging to a family of perpetual wanderers. In 1954, McCarthyism forced my father to flee the United States (just as fascism had previously forced him to leave Argentina), and the family trailed him to a Chile where everyone spoke Spanish, that dreaded, despicable, alien tongue. It took me a while, but ever so slowly I fell in love with the land and the language and, ultimately, with one woman, Angélica. I also fell in love with the peaceful revolution led by Salvador Allende, so by the time he had won the elections in 1970, I had reappropriated the word *forever*: I would live and die forever, *para siempre,* in that one country of my dreams, I would see social justice in all of Latin America, I would not need to speak or write in English, not ever again, a language that my febrile radical brain identified with imperialism and U.S. domination. Nor did my attitude change when the 1973 coup ferociously descended upon me and my people, and I was flung into an exile I did not desire. I would stay faithful to my Spanish, I told myself. I would return in glory, *en gloria y majestad,* to Chile. Along with my people, *mi pueblo,* I would emerge, we would emerge, from the shadows. *Saldremos de las sombras.*

In July of 1990, when the dictatorship ended and I returned to Chile in what I thought was a final homecoming, it seemed that my prophecies had come true, it seemed that my exile was over.

History had other plans. Six months after our arrival in Santiago, Angélica and I packed our bags one more time and headed north—and here I am, twenty years later, sitting in my study in North Carolina, writing this introduction in English, far from Chile and far from the young exile who swore that some things are forever, here I am, living in a land defined by a second September 11 and yet another act of terror. It is true that when I finish this text I will immediately turn to my Spanish, as I did that day when I was informed that I had been murdered in Chile. And it is true that I still predict that we shall, our species shall, someday emerge from the shadows.

And yet this is not the future I had imagined, this separation from my community, this mongrel heretic of language that I have become, this insurgent nomad of the earth who writes these words.

How did it come to pass? How did the exile I had been so intent on renouncing forge me into someone who could not find a way home? Why did my country not respond as I expected to my love affair with it? Who was to blame? Was it that land, the world, me? And how is it that I became a bridge for the multiple Americas so often at war in the outside world of murderous nations and forbidden borders? Was it necessary and even inevitable that I should end up thus, my Spanish and my English making love to each other inside me after so many years of fighting for my throat? Was there a deeper meaning or message in the maelstrom of my dislocations, had I learned a lesson worth sharing with others during these intense journeys to the South and then to the North, this back-and-forth of body and mind into which my existence has been transformed?

It has taken me many years to grapple with those baffling questions about my own identity. I made a first attempt in the mid-1990s to come to terms with the reasons why I always set out resolutely in one direction and ended up, over and over, taking a different, even opposite path, as if the sadistic demons of history were bent on playing with me and my expectations. But that book, *Heading South, Looking North,* only took me and the reader up to the coup of 1973 that forced me to flee Chile; it left the author dangling and about to start his exile, left many questions unanswered. I may have avoided tackling the decades of death and tenderness and solidarity and despair that followed because they were too painful to deal with, maybe I needed time to understand the hidden traumas of separation, the revelations that come to us when — as in the course of a terminal illness — we suddenly see in an entirely new light family and friends, light and darkness, betrayal and loyalty, power and responsibility.

Or maybe I was wise enough to know that we do not choose our books but, rather, they choose us, they are aware, sitting or seething quietly in the recesses of the heart, that their moment of birth will arrive only when the midwife, the birth canal, the many strands of creative copulation come together, demand relief and recognition and air. Waiting for the seeds that have been planted who knows how long ago — such a tired metaphor and yet so true — to ripen into words.

When was that moment for me? When did *Feeding on Dreams* start to swim towards these words I now write?

It must have been a vague day in 2006 when I accepted the suggestion of Peter Raymont, the renowned documentarian from Canada, that he film the story of my life in the three countries that had defined it. Or perhaps it was during the subsequent journey to the Argentina of my birth, to the Chile I had returned to and then left for good, to the New York I had never wanted to leave as a child, maybe it was while I retraced and filmed my steps that some earthquake of language began to shudder me into the need to make final sense of my existence, to attempt to resolve in my literature what my life had denied me, craft some illusion of home and stability, some sort of pattern behind and inside and beyond my incessant drifting from language to language and land to land, the fate of so many in our times, so many in our forgotten past.

The project was given more urgency, I believe, by the events of 9/11 in the United States, that strange and terrible Tuesday when my life was torn asunder again as death surged from the sky, the second September 11 of desolation I had experienced. More urgency because that day confronted the citizens of the land where I lived, where I had made my home with Angélica and our two sons, where my granddaughters were born, it forced the United States, and therefore the world as well, to confront the major questions about violence and forgiveness, memory and justice, tolerance and terror, that I had been working through with my fellow Chileans and fellow exiles for a good part of my life. More urgency because we live in times when, in some twisted sense, we are all exiles, all of us like a motherless child a long way from home, times when we are threatened with annihilation if we do not find and celebrate the refuge of a common humanity, as I believe I did during my decades of loss and resurrection.

So here is my story.

As near to an understanding as I can achieve.

It is, above all else, I suppose, the story of how I tried to defeat death in the late twentieth century, what I would like to remain behind when it comes for me and I won't be around to state that, alas, the reports of my death have not, after all, been that greatly exaggerated.

PART 1

ARRIVALS

You will be separated from yourself and yet be alive.

— Ovid, *The Metamorphoses*

Fragment from the Diary of My Return to Chile in 1990

JULY 21

All I ever wanted was to come home.

That's all I wanted, day and night and hour by hour, year after year, my one devouring obsession, to return to Chile. Ever since I was forced into exile in early December of 1973, though perhaps even before that, perhaps from the moment of the coup against Salvador Allende, something whispered to me, as the bombs fell on the presidential palace where I should have died and did not, perhaps from the moment I escaped death as a prelude to escaping the land where death suddenly reigned, ever since then, to return, *volver, voy a volver,* that's what I swore I'd do, find a way back and never have to leave again.

To see the mountains in the morning, every morning, just like now, on this first dawn of my final return, what more can I ask for as I begin to write this diary of my homecoming, these words with which I greet myself this dawn. There for me, *los Andes,* my sentinels, telling me that it's over, my exile is finally over.

The mountains — the first thing I remarked on when I arrived in Santiago, barely more than a child of twelve, speaking not a word of Spanish. One of the first words I must have learned, I must have asked my parents what they were called, I — wait, wait, you're making this up. You did not want to speak a word of Spanish, but understood most of what was said in that language, had heard the term *montañas* many times, so

let's honor this *retorno* by not overly exaggerating things in
this journal, Ariel. Prove you've really left exile behind. Because out there, in that merciless world where nobody knew
who you were, you could invent your past, try on a past as others try on neckties. There's a reason why the Greeks called exiles liars, pretenders, manipulators of facts, because who could
corroborate their boasts about exploits or wealth back home,
who knew their mother and their grandmother and their an-
cestral threshold? But here in Chile, where a community of
welcoming hands and prying eyes awaits you, well, no more
twisting the past for effect, what do you say?

This much, then, is true: it was not love at first sight between me and the mountains.

My first impression of the Andes, back in 1954 when I was
forced to follow my persecuted father and my loyal mother,
forced to leave the New York I adored to settle in a country I
did not care to know, far from my baseball and my musicals,
far from my friends and our apartment on Riverside Drive
from where I could watch the slim-legged girls of Barnard
playing tennis, far from my teachers at Dalton on East 89th
Street, far from everything familiar, what struck me at the beginning of that childhood banishment to the Andes was . . . I
guess the right word would be *terrifying*, the mountains struck
terror into me. Those unconquerable slopes, that vastness
blotting out half the sky, overwhelmed the small beat of my
heart with a sense of suffocation, isolated me from the world,
told me that I was trapped in this foreign place without any
possible reprieve.

But it was not long before I realized that the immensity of
that mountain range contained a proposal of shelter, an almost
maternal intimacy for those who resided under its peaks. Because it was enough to glance up to know where the East lay,
and the South, and therefore the North I pined for: nobody in
Chile could be disoriented. Maybe that's why, for a boy who

had been cast out of the Argentine city of his birth as an infant and then had to mourn the American city of his childhood, to abide in the valley of Santiago became a way of living with the gods at his back, guarding the entrance to paradise.

And as he allowed the country to seduce him, it had indeed been like paradise for that boy who turned into an adolescent and then a young man, except that it was too visibly an inferno for too many of its inhabitants, the blight of poverty and injustice plaguing that land so full of delights. I was fortunate, nonetheless, to come of age at a time when those inhabitants were ripe for the promise that was Chile, at a time when a majority demanded a taste, and more than a taste, of heaven. I was blessed to grow into manhood at a time of revolution. Not ordinary, this revolution inaugurated by Allende, when he was elected president in 1970: one that was peaceful and democratic and did not believe that you must kill your adversaries in order to create a world of justice and freedom for all. One that was as vast as the mountain range that presided over our three years of joy and exultation, because nothing can compare to inhabiting the giddy center of history, nothing compares to being set gently on fire with the knowledge that the world does not have to be the way you found it, the certainty that everything wrong and unfair and ugly in the world can be changed, nothing can compare to the conviction that we the people, *el pueblo,* are the sovereign masters of our fate, three years marching with millions of others towards liberation under the gaze of those mountains, as if they could guarantee that nothing could corrode our dreams, as if the horizon were there for the taking and death did not exist.

And then the mountains vanished, swallowed up by exile. There was a void in the scenery when we most needed some substitute womb to crawl back into, at the very moment when all vestiges of permanence were being blown away, the country of our love decomposing under the double assault of Pinochet's dictatorship and the less violent passage of time with its grind of inevitable blurring and misremembering. All that was left: a brooding sense of emptiness, the compass of existence abolished, the cardinal directions erased by some fiend.

Unmoored. Desperate. Queasy.

As soon as I arrived in Buenos Aires from Chile, scarcely months after the coup, there, in the flattest city on earth surrounded by thousands of miles of pampa, I could feel the mountains calling from inside me, crouched in some unforgiving, unforgetting network of cells in the brain, waiting for a pretext to materialize, waiting for me inside my neurons and back here, waiting for this morning in Santiago at the end of July of 1990, when I would be able to drink in their stability and not feel everything *tambalearse,* disintegrate around me.

As my days of banishment turned into months, and the months into years, I learned to control the vertigo of dislodgment, pretended that I could live without the mountains — that's what exile does to you, like the dwindling away of a love you held dearest and swore you'd never survive, and then one day you find yourself drinking wine with someone else, you find yourself walking down a foreign boulevard and there is not a twinge of regret, you can live away from the mountains, you tell yourself. Except you can't. Except you watch what happens to your son — the mountains also began to disappear from my Rodrigo's drawings. No small event, this. Ask any child from Chile to sketch something, anything at all. Before any human figure, a cloud, a tree, they'll fill the upper space with an array of jagged peaks — so when the zigzag mountaintops began to wane from the pictures of our exiled sons and daughters, we took it as a warning from the very earth of Chile to beware, *cuidado,* the country was receding from their eyes and perhaps from ours.

And then a child is born in exile, as our Joaquín was, in an extremely horizontal Holland, and there was not a hill in any of his drawings, and we didn't even attempt to nudge him into smuggling a mountain into his picture, force him to retain our memories or enact our desires. In this, as in so many things, Angélica has been wise, and insists on keeping our youngest son safe from the adult manias and quests that I injected, exhorted, into Rodrigo. Now that eleven-year-old Joaquín has returned with us to the Chile where he was not born, his dreams and his drawings will slowly begin to overflow with undulations.

We'll see how long this conditioning takes. Joaquín is not enthusiastic about living here or leaving his friends behind. He's safe from the dictatorship, we've only come back permanently now in 1990, now that Pinochet no longer rules Chile; but we cannot keep Joaquín safe from the aftermath of that dictatorship, he isn't safe from the potential ravages of this return. Nor safe from repeating his father's story. I was more or less his age when I was uprooted from the States and dumped in an alien land. But he has cousins, he has visited often, he speaks enough Spanish to get by, he has an elder brother to receive and coddle him, he'll work it out, I'm sure he will soon find the mountains to be his best friend, nothing to worry about.

Am I speaking to him or to myself?

Am I the one who is worried, needs comforting? Under all this early-morning euphoria, do I sense an unease that will not melt away? Have I fixed on the immobile constancy of the mountains because I fear that the man who looks up at them now has been hammered and reforged in ways that will make this return arduous? Is it possible that these seventeen intervening years have changed me and changed the country in ways that are irreparable? Would the people of Chile, the *pueblo* I so ardently defend, understand these hesitant words meant for my own consumption?

The mountains are still here, but so is Pinochet, so is his influence.

He was compelled to withdraw from the presidency two years after being beaten in the 1988 plebiscite, but he is still commander in chief of the army for the next eight years, and from that impregnable position he issues threats and throws tantrums, reiterating that if one of his men is touched, the state of law, *el Estado de Derecho,* is over — in other words, he is more than willing to unleash another September 11, 1973. Just as decisive as the fear he continues to instill in a traumatized population are the bonds with which he controls the transition to democracy headed by Aylwin, our newly elected president. I use the word *bonds* advisedly. Pinochet has boasted — mimicking Franco — that he has left "*todo atado y bien atado,*" everything tied down, well tied down. His constitution, fraudulently approved in 1980, warrants that our democracy will be *protegida* (but protected against whom? against the people, against an excess of democracy) and *tutelada* (as if we were children requiring tutors and supervision). Consequently reforms become unattainable because all the institutions have been bolted down to their places: the Constitutional Tribunal, the Council of State, the Supreme Court, the armed forces, each one an authoritarian enclave stacked with supporters of the former regime. Even the Parliament cannot function without the consent of the Pinochetistas, because the senators he named before leaving office can veto any legislation that menaces his legacy. And his followers dominate the economy, the media, the university hierarchies. Tied down, well tied down. Bound by previous legislation: the *Leyes de Amarre. Amarrar:* to fasten, clinch, rope in.

What I find fascinating — and chilling — about these words inflicted on post-Pinochet Chile is how they insinuate and portend a subtext of bondage and repression. Chile is being treated as if it were a prisoner tied — *amarrado* — to a chair or a cot in some dank basement, each limb lashed in, each eye-glimmer of protest slammed shut, reminding me of the fate of the *imbunche,* a legendary creature in Chilean mythology, a baby stolen from its parents by underground demons, all of its orifices sewn up, blind, deaf, mouthless, a defective child subject to experiments by its captors.

I am obviously exaggerating. Pinochet is not omnipotent and we will without doubt find ways to limit his authority and sway in the years to come. And yet, that I have so noticeably slipped into images of cruelty and coercion, that I have attached myself to metaphors of torment and deformity, is a disquieting sign of how I have been penetrated by the terror that has corrupted the country. Maybe I'm the one tied up, with other, less visible, less obvious ropes. Maybe I'm the one who is *atado y bien atado.* Ropes provided by history, by exile, by what I have had to do in order to survive, so that we all could come home. Is rejoicing at the permanence of the Andes one more fantasy? Is it possible to enjoy every mountain peak and every crepuscular sunset and still no longer fit in, no longer belong here anymore?

I will soon find out.

•

IT WASN'T as if I had not been forewarned.

That's what I say to that alter ego of mine, the man who wrote the diary in 1990, that's what I can realize now, twenty years later.

As I remember the evening when I had dinner at the Smithsonian, in Washington, D.C., with Bruno Bettelheim, the Austrian author and child psychologist, sometime in the early eighties.

"How long have you been away from Chile?" he asked. He had been in the States since 1939, after having escaped Hitler's Germany and eleven months in Dachau and Buchenwald. When I told him that it had been seven years at that point, he nodded and smiled. He was not a jolly man but I remember that he smiled. "Good," he said. "If you are gone for fifteen years, you will not return. Even if you return, you will not return."

Even if you return, you will not return.

My answer to Bettelheim that night was to recount a story, a possibly alternative scenario, that had been confided to me in Buenos Aires by another author, the preeminent novelist Augusto Roa Bastos, when I had gone to see him in late 1973. Demoralized and defeated and at the start of an exile that had no clear end in sight, I sought out Roa as a real expert on diaspora and bereavement: in 1946 he'd fled the brutal dictatorship in his native Paraguay and still couldn't go back to the rivers that had inspired him. He was one of untold numbers of his compatriots living abroad — in fact, one-third of the country's people had left, the equivalent of one hundred million Americans forced from their land. Did he have any advice?

The Guaraní Indians, he said to me, believed that exile is a form of death, that he who is banished journeys into the country of the dead. Beware of this journey, that was his advice. In societies that Westerners in their ignorance consider primitive, the worst insult is to call someone an orphan, because it means you have lost your community, what gave daily meaning to your life. Beware of the orphan you may become, that ostracism can turn you into, beware of becoming a ghost.

And what happens when you return?

He looked at me enigmatically. "If you return, you mean, *if* you return."

"Yes. What happens if you return?"

"The Guaraní celebrate the return of an exile as if it were a resurrection."

I held on fiercely to that story of the Guaraní Indians through the seventeen years that followed. And when I finally did return, in 1990, I used the mountains as proof that Bettelheim was wrong, that exile does not mark you forever, that it can be a mere parenthesis. I dismissed the mayhem of doubts that accompanied me — what if Bettelheim had been right, what if my real future lay abroad, what if Chile had become the country of death rather than

the land of life, a country dripping with fear and ropes and inhabitants sewn up like an *imbunche*, like a deaf-mute child, what if it
made no sense to return?

Those dark thoughts were no way to begin a homecoming.

I wagered back then that the mountains were telling me the
truth. I celebrated my return to Chile as a resurrection, I trusted
that I would not be, could not be, an orphan forever.

ARGENTINA, 1973. That's when and where it all started, all the
dilemmas seeping out, the future already written by history for
me when I was least ready, most lost, the decisions I made while
hardly understanding what I was doing; all this leading me to this
day in 2010 when I try to make sense of the past.

No sooner had I landed at the Buenos Aires airport in early
December of 1973 than the Argentine police shattered the fantasy that I would be able to remain there for as long as I wanted.
The interrogation by the security forces lasted for hours, until
they let me go with a warning: I'd damn well better behave myself. Behave myself? On the contrary, I expected the right-wing
Peronista government to aid and abet my revolutionary activities
by providing me with a dark blue passport that would allow me
to crisscross the globe working for the Resistance protected by *la
flamante República Argentina*. I might have renounced my Argen-
tinian birthplace upon becoming a citizen of Chile, but it could
not, by law, renounce me — that was the plan I had hatched in Santiago once I realized I was bound for exile. It was the prospect of
that illusory passport that had allowed me to refuse the status of
refugee, declining the offer from the United Nations functionaries who came to interview those who had sought asylum in the
Argentine embassy. Hunted down and humiliated, I wanted to
hold on to one straw of self-respect as the waters rose to engulf
me. I was determined to cast myself in the role of heroic exile, a
Byronic avatar roaming the continents in search of justice and
beauty, rather than another faceless victim among a flood of anon-

ymous millions fleeing the planetary carnage. Sheltering my battered ego, I pretended I could exercise some control over who I was and where my body could travel.

And in spite of my less than warm welcome to Argentina, I was still under that illusion when, the morning after my arrival, I set out, full of love and confusion, to meet the leaders of the MAPU, the revolutionary party I belonged to. Armed with my chimerical liberty of movement, shielded by the Argentine passport that I did not yet possess and would receive only after arduous months of wrangling, and then only thanks to the last of my father's faltering contacts in the government, I offered the three men heading the party in exile jurisdiction over my body and even my soul. It's hard to believe now that I would put my life and that of my family in the hands of men no older (and often younger) than my thirty-one years, who were just as mystified as I was by the novel situation we were all living in. But those were times of war and I saw myself as primarily that, a warrior, a militant. A word that comes from *miles militis,* Latin for soldier. The image I wanted to project: a soldier of the antifascist struggle. Someone who finds meaning inside a hierarchy of command and consents to becoming a pawn in a strategy decreed by the collective wisdom of superior minds.

Not the only key encounter that Buenos Aires had prepared for me. I had been summoned that first morning of my exile by Jacobo Timerman, the remarkable journalist who ran *La Opinión.* That Argentine newspaper had awarded me a prize for a wild, experimental novel I had written about the Chilean revolution that had just been published in Buenos Aires. Maybe he wanted to help me promote the book?

That wasn't what was on his mind. "I have a proposition for you," Jacobo said. "Work. Four articles on what occurred inside the Argentine embassy over the last months. A thousand refugees, from all over Latin America. Interesting things, strange things must have happened."

It certainly would have constituted a heartbreaking chronicle of a revolution gone down in flames and its champions left marooned, listening to death rampaging just outside the walls. But I said no thanks, no, I could only tell that tale if I revealed the whole truth, and this wasn't the moment to expose our disarray and dirty linen to the world, I'd have to delve into betrayal and pettiness and cowardice. Lost amid the anecdotes about the man who had gone to the embassy to escape his wife and the insults traded by three factions of Bolivian revolutionaries would be the story of those who kept their passion and faith alive. My articles would distract from the fight against the junta and the image of victory we needed to nurture if we were to win.

"I respect your decision," Jacobo said, "but I think it's always necessary to tell the truth, no matter the consequences, and as for the time, *siempre es ahora,* it's always now. We shouldn't subordinate our writing to any politically expedient cause, no matter how noble it may appear."

I'd remember his words a few years later when he published the names of the Disappeared of Argentina on the front page of his paper, I'd remember Timerman when they arrested and tortured him for daring to do so, and I'd remember his words again when he wrote his searing book on Israel, the country that had received him once he'd been released from that ordeal.

And I remembered those words and used them, in fact, when I returned to Chile in 1990 and wrote my play *Death and the Maiden* in spite of those who suggested that perhaps I should consider not upsetting the apple cart, that some apples are indeed forbidden to us, that we need to be conscientious and not provoke the military, not stir up the flames of retribution. I said to myself then, the time is always now.

But that was not the sole lesson about writing and responsibility posited to me that day in Buenos Aires. As I was leaving *La Opinión* later that morning, after meeting up with Argentine friends who worked there, as I headed for the elevator, I bumped into a Chilean writer, whose last name I can't recall but whose glee I still can't forget.

"I just came from meeting with Timerman," he said, pushing the button for the lobby, "and I'm going to write an article for *La Opinión* about the death of Victor Jara. A thousand dollars," he chortled, "that's what he's paying. The whole world is going to know that they cut Victor's hands off with an ax!"

"But they didn't," I bristled, "when they murdered him. I heard that they beat him, broke his hands maybe, but nobody cut his hands off with an ax."

He went ahead and wrote the piece, and the legend of the invincible Victor Jara singing in the stadium with his stumps spurting blood still persists decades later.

So there it was, greeting me from the very beginning, the multiple quandaries of every writer in exile, of every writer trapped in the war years, of every writer as he or she searches for the many elusive faces of the truth in a world defiled with lies.

That first day in Buenos Aires was giving me a taste of what it would be like to navigate the temptations and pitfalls that open when you're far from your land. Buenos Aires was where it first dawned on me how vulnerable a foreigner can feel, what befalls you when all the shelters you take for granted dissolve. Buenos Aires was where I discovered that Kafka was a realist, as I shuffled from one waiting room to the next, seeking someone, anyone, please, who could offer a way to obtain a passport before the death squads came for me. Because it was clear that a massacre was about to occur, a repression just as terrible as Chile's about to descend upon the Argentine revolutionaries, and anyone in the slightest way associated with them, *they are going to kill you, all of you, don't you understand? They're going to hunt you down as if you were dogs, worse than if you were dogs, because dogs will have rights that you won't dream of*—and my Argentine friends responding what I had responded to in Chile up until the coup, as adamant and deaf as I was back then, reassuring me that such things did not take place in their country, the sort of drastic interruption of democracy that had destroyed Chile, surely I was exaggerating, not to panic, not to worry, *we're not going to leave our land like you had to leave yours, Ariel.* It was in Buenos Aires that I first realized the futility of being a prophet in a foreign land if no one understood your garbled tongue, it was in Buenos Aires where those lessons began to be learned, those lessons and so many more.

So it made sense, when we were planning the 2006 journey to film the story of my life, to pick Buenos Aires as our first stop. The film's director, Peter Raymont, wanted shots of me visiting the places where I had spent my infancy, wanted to record me as I went to speak to my sister Eleonora and my cousin Leonardo and above all to capture me leafing through family albums seeking photographs and documents that would attest to my family's lineage of displacement.

As for me, I had unfinished business in Buenos Aires. My thirty-nine-year-old son Rodrigo, who was coming along as the location scout and associate producer of the documentary, had been insisting for years that there was a debt I had to pay in the city where I'd been born.

Just before leaving Buenos Aires in February of 1974, I had stopped to say goodbye to Baba Pizzi, my favorite grandparent, the one who knew most about exile, since she had been forced to leave her native Odessa not once, but twice. Not the only bonds that joined us. She'd been a quasi-Bolshevik, Trotsky's interpreter at Brest-Litovsk. She had heard Trotsky say that he was willing to concede everything to save the revolution, everything except the revolution itself. And we shared a love of literature. She'd been the first Spanish translator of *Anna Karenina,* a journalist and author of stories for children.

What I loved most was her fierce enjoyment of life, that survivor of odysseys and catastrophes, a woman who had gone through the First World War and the Communist insurrection and the Russian civil war and had managed to escape alive with her son,

my father, not a scratch on her body, and then, in Argentina, decades later, walking to the bus stop on a day filled with the skirmishes that accompanied who knows what general strike, my
Baba Pizzi had heard a shot and looked down and her index finger
was gone, just like that, a fascinating item for her little grandson,
to rub my own fingers against the shiny round stump. Our relationship was made possible because she spoke almost perfect English (and Russian, French, Spanish, German) and I was a monolingual imp who spoke only the language acquired in New York.
My Baba Pizzi provided the one thread back into the past, a link to
the ancestors, all those generations coupling through history, and
to the future as well, because she adored Angélica, her *nietecita*.
And on our honeymoon, in 1966, when Angélica and I hitchhiked
from Santiago to Buenos Aires and Montevideo, my grandmother
had taken care of the newlyweds, only endangering us all when, in
a taxi in Uruguay, she had berated and beaten with her umbrella
a driver who, she discovered, was a White Russian and therefore
deemed personally responsible for her penury during the siege of
Odessa. How could it be that I would not share her last years? And
yet, the right thing to do, the only thing to do: a few days after we
left Argentina in 1974, a death squad descended on my Baba Pizzi's
apartment.

She told me the story in a letter that I received in Paris, and I
called her to hear her bird-like voice, with faint traces of Russian
still there, assuring me that she was fine, she knew how to deal
with that sort of scum.

A few years later, another telephone call, this time from my father in Buenos Aires, announcing that my Baba Pizzi had passed
away. Though in a sense, exile had killed her before she died, she
had disappeared day by day from our lives once we had been
forced to leave Argentina, so we had already mourned her absence
before it became final.

Perhaps that's where I wished to keep her death, conveniently
suspended beyond time and space, perhaps I did not wish to con-

front the corporeality of that loss, perhaps I had been burdened with more pain than I knew how to cope with, more deaths than I knew where to store. There are many reasons for the fact that when the disastrous invasion of the Malvinas by the Argentine military led to democracy being recovered in 1987, and I was able to return to Buenos Aires for the first time since our getaway, I did not visit my Baba Pizzi's grave. Not then, and not during any subsequent visits to Argentina.

"This can't be," Rodrigo said to me before we left for the filming in 2006. He was particularly puzzled by what it meant that the ashes of my dad and mom, his grandparents, were in two urns in my sister's tiny Buenos Aires apartment, waiting for us to decide what to do with them, whether to take them back with us to the United States or scatter them in the Río de la Plata. Meanwhile, their remains had no final resting place, whereas at least Baba Pizzi was in a cemetery. "I think it's really strange that you've never gone to pay your respects to her," Rodrigo said, "tell her about Isabella and Catalina, that now that you have grandchildren you understand how she must have loved you, how she must have missed you when you left as a baby and then again as a young revolutionary to save your own life."

And so one morning, trailed by our fly-on-the-wall film crew, we set out for La Chacarita, where a certificate specified where Baba Pizzi and my grandfather Dieda were buried.

Rodrigo and I were in for a rude awakening.

She wasn't there.

The number was right, the *pabellón* was right, but no, some stranger was resting in my grandmother's alcove — all those names engraved in the wide wall, with flowers and other signs that they were cared for, a hand had come to shine the marble, leave a message, name after name after name and among them none that spelled out the words *Raissa Libov de Dorfman*.

A measureless sorrow began to well up from within. As long as in some recess of my existence I knew that Baba Pizzi was safely interred, there was no need to exhume her memory or realize that she was really dead, but now . . . Where was she?

A nice woman on the cemetery's staff explained that nobody had paid for that niche in years, and so my grandparents had been taken to the *fosa común,* the common ground where paupers are buried, where those neglected and forgotten, the endless armies of the disinterred, all end up, where "we all end up," she said softly and wisely, "one day or another everybody ends up there."

Now I was truly shaken. I thought of the hands that were not mine and not Rodrigo's carting my Baba Pizzi away, violating her peace, with no ceremony and no ritual, no words of farewell or hello, how could I have facilitated this desecration, however un-wittingly?

And then, some comfort.

As Rodrigo and I walked towards that common ground, I saw men working on the gardens and amid the tombs. They were the ones, those men, or some brother or colleague of theirs, they were the companions of the dead, they lived among the dead and lived from what the dead bequeathed to them, their children ate thanks to the seeds and rot of the cemetery, they were the right ones to have carried my Baba Pizzi on her final journey. They had passed her ashes day after day, had kept the marauders and the wild cats away, had tended to the ground and stones and plants surround-ing her while I was desperately surviving the cruelty of Paris and the distances of Amsterdam and the difficulties of Washington, D.C., and the military perils of Santiago de Chile, they had taken care of my loved ones when I had been unable to do so, they had the right to spirit her away into her last night, they must have given her some love and tenderness born of familiarity and duty as they let her dust fall to earth.

When I came to the ground itself, with its verdant pastures and majestic trees and a wall covered with graffiti where relatives like me had come to write their homage, when Rodrigo and I came to a particularly noble-looking tree, it all began to make sense.

Sobbing in each other's arms, sobbing for our Baba Pizzi and our
Dieda and my forever parents and all the rest of the endless dead,
but above all for ourselves, we stood in front of one tree that chose
us, called to us — here's where she was, under these leaves, and Rodrigo handed me photographs of his girls and he had bought flowers and we placed one flower after the other next to the roots of
that giant friend with its boughs reaching up into the sky, each
flower accompanied by the name of a surviving member of our
family, this is for Angélica and this is for Joaquín and this is for
Isabella and for Catalina and for Rodrigo's wife, Melissa, this is
from me, and then I spoke to the tree and asked it to take care of
my Baba Pizzi, reminded the tree that my grandmother had been
a writer herself and that as far as I was concerned this meant that
the earth should treat her with particular care, be sure she was not
too lonely.

Was that peace I found in La Chacarita just another delusion? One more way of assuaging the demons of exile and making believe I was not evermore damaged, that they had not interposed themselves between me and my Baba Pizzi? Was it a
delusion that the links in the long chain of ancestry had not been
broken?

If it's a delusion, it is a sweet and significant one.

An old woman in Chile — not as old as my Baba Pizzi, but old
enough — once told me a story. Every Sunday, this widow, whose
name was Sarah, visited her husband at the cemetery, she said, at
about the same time another woman came to deposit flowers for
her dead son, and they kept each other company in their eternity
of loss, and so it had been week after week. Then one Sunday — it
was the first Sunday after the military coup — that other woman
had not been there when Sarah went to the cemetery, nor did she
appear again the next Sunday or any other Sunday after that. So
Sarah started to care for both graves, she bought flowers for her
own husband as well as for the unknown child who lay close by
and who apparently no longer had a mother. That's the way it is,

the way it has to be, *así tiene que ser,* Sarah said to me, because when I am gone someone will take care of me and clean my husband's grave and also that boy's resting place, and when there is no one left, the wind will do its job, the wind and the rain will do what they must, and it will be all right, there will always be someone who remembers.

My consolation, then: there will always be a tree in Buenos Aires that remembers. In Buenos Aires, where so many of my journeys had their start.

Fragment from the Diary of My Return to Chile in 1990

JULY 23

When I returned to Santiago in 1983 on the first of many visits, I just couldn't stop writing. For the next seven years, maddened by the desire to compensate for what I had missed during my exile, I ravenously drank the boundless ocean of everyday Chile and just as avidly poured out an ocean of words, tried to transfer the country into a fury of meaningfulness, interposed the sin of impatience between myself and the land, myself and the people. A pattern of profligacy and haste that also pervaded the way I've lived my banishment, a potential deafness to what reality may be muttering as I rushed from new project to newer project, from article to story, from story to essay and essay to poem, this play and that novel and the next novel, on and on, this op-ed and that speech, goaded on by the faith that my incessant creativity was a way to prove that the dictator had not crushed my spirit.

Well, Pinochet is gone, and democratic Chile in 1990 now offers me the balm of peace, an interval when I can gather my bearings and catch my breath. And so I've taken a solemn vow of silence, sworn that I won't write a thing, not a word, during these upcoming six months here, at least nothing destined for an audience. Of course I've sworn many things in my life and more often than not found myself forced to alter my plans. I swore the revolution was unstoppable, I swore I'd rather die than go into exile, I swore it would take us a year at most to oust the generals and return to Chile, but I think this vow of silence should be easier to keep.

Maybe this chronicle of a return foretold is a sort of compromise — a way to keep the words rolling cautiously, but for my eyes and no one else's, a way of taking stock of where I am at this critical juncture and how I got here. A guarantee that I'll spend this period unobtrusively, avoiding controversy and the public arena, sedately edging into the country. Perhaps this is a way of addressing my compatriots from a distance akin to that of exile, a way of practicing for the book this diary may someday become. For now, I'm sending these messages read only by me to all those who may not be ready for them quite yet.

Because it really does seem that the time has come to yield to the muses of slowness and acquiescence and wait for Chile to speak to me, let my country tell me what it wants from my hands and mouth and eyes.

•

IT'S EASY TO smile at my vow twenty years later, now that I know it was not to be. That I could no more remain quiet and withdrawn on that return in 1990 than I could stop breathing.

Fraught with irony even then, that forceful declaration of self-restraint, given that it was precisely silence that had almost destroyed me during the first two and a half years away from Chile, a plague festering in my throat that I could not comprehend and felt that I did not deserve.

I had not, of course, been able to write one phrase during the three weeks I had been hunted down by the military after the September 11 coup, and circumstances were hardly more propitious while I was crammed in with nine hundred other disposable asylum seekers at the Argentine embassy in Santiago, but I predicted that as soon as the junta allowed me to cross over the cordillera to Argentina, I would find my voice. Emerging from defeat, weighed down by the memory of those who had died in my place, accusing myself of weakness for having accepted the order to leave, I fought off survivor's guilt by trusting that I had been spared in order to tell the story, keep our resistance and memory alive in language.

What awaited me instead, in Buenos Aires and beyond, was a blank page.

I would stare at the quiet, recalcitrant typewriter for hours, my fingers dysfunctional, unable to indulge in the slightest flight of imagination. I shook off that numbness to do political work. I clacked out letters to friends in Chile and lobbied powerful people abroad, badgered journalists to cover our specific cataclysm among so many crying for attention, composed memos to the banished leaders of the Resistance. I contacted foreign artists and musicians and authors to see if we could send money to our distraught colleagues in Santiago and Valparaíso and Valdivia. I compulsively spelled out the names of political prisoners and the names of the executed, my life was full of names and exhortations, all of it urgent and supremely *eficaz,* hack work that needed to be done to save lives, my small way of helping to save the country. Work that would not, could not, save me.

And it was salvation I needed, the same salvation that literature had afforded me from the age of nine when I had first discovered its wonders, first used writing as a bulwark against solitude, a sanctuary against chaos, I said to myself when I had been forced to leave New York for Chile so many years ago, something I can carry with me. I comforted myself with the thought when again I had to leave the country I had adopted, assuring my battered self when I left the Chile now belonging to Pinochet, the one item I could transport with me like a toothbrush, smuggle invisibly out of the country, past the border guards and the censors, as much a part of me as my ankles or my sex.

I had not been valiant enough to die next to Allende, but I would at least become the guardian of the words forbidden back home, the words that persist only in the shadows back home — I can hear myself now, that lyrical earnestness! — I would turn the curse of distance, its exacting punishment, into the blessing of creativity, connecting readers to a past that was being fractured like the legs of a bride on her wedding night, every last bone in her body broken on the night she should have wed. I would, in the words of one of my favorite songs of struggle, overcome. It was a matter of opening the faucet of my inventiveness and letting the words flow.

And I had opened the faucet as soon as I had escaped from Chile, and not a drop of water, not an inspired, inspiring word dribbled out, no story at all, nothing, no muse was saddened by my plight.

My literature had dried up.

Only once did something come to me, like a disease in the night. Just before dawn, in a Paris hotel in May of 1974, eight months after the military takeover, I awoke and felt the urge, grasped five words like an imprecation, stood up in the darkness, skirted my way to the bathroom, closed the door behind me, and flicked on a desultory lightbulb hanging like a corpse from a shredded ceiling.

Softly, gingerly, I lowered the lid of the toilet bowl, sat down on it, and placed the green Olivetti on my knees, and there in the hotel on la Rue Blomet, I wrote out what had been soiling my mind, hiding in my mind, rapidly pounded out those words in Spanish that I could finally formulate. The sound of each key striking the paper seemed oddly disruptive, that tapping sound which had, for as long as I could remember, been normal and friendly, a sound suddenly harsh, suddenly undesirable. It was five in the morning and darkness surrounded me, so peaceful that I could hear Angélica and Rodrigo in the next room, the hush of their breath in the night just beyond the closed bathroom door.

So the racket hadn't awoken my wife and son or anybody else. That was all I needed, a cohort of guests hurling catcalls up through the inner courtyard of the hotel where in the daytime tired immigrant women hung their laundry from unwashed windows as they warbled to each other in Arabic or Portuguese or French. I could picture them hurrying up the stairs to the fourth floor and banging on the door, demanding that I cease and desist. I almost wanted that to happen, for somebody to stop me from confronting those words I had snapped out of the depths of my despair, the moment when, adrift in an alien city, I began to meet the man I had become, trying to understand what I was doing there in the loneliness of yet another strange neighborhood, besieged by the predawn certainty of ruination and the acrid smell of walls stained from too much frying, washed by the tides that had deposited me on the shipwreck of this toilet and those five words.

Cómo pudimos habernos equivocado tanto?

How could we have been so blind?

That's what I wrote from the deepest pit of myself, what crawled back at me from inside the whiteness of the paper, hardly visible under the dim light of that tiny bathroom, those words nailing me, asking how could we have been so wrong, how could we have made so many mistakes?

And then I let myself go, let loose an anguished screed of recriminations. This was no ordinary failure. We had promised our people paradise and they had reaped, we had reaped, the winds of hell. *We* had promised? Me, me, me. I had promised, I had been blind to what was coming, I had proclaimed victory right around the corner, I had marched through the streets of Santiago vociferating hymns of liberation, I had waltzed through three years of rebellion, I had predicted a glorious future when not one child would go hungry, not one peasant would be landless, not one miner ever again exploited, not one natural resource left in the hands of foreign interests, not one woman bereft of a tomorrow, not one set of eyes unable to read, not one hand afraid of reaching out to the hand nearby.

But it wasn't a flowering of democratic socialism we were bequeathing to our children. Look at the promised land my Rodrigo would inherit if he ever returned from a banishment he had not chosen, look at this aberration. Forty years after the end of the Second World War, an underdeveloped version of fascism. Centuries after the last heretic was burned at the stake, the reign of a new inquisition. A throwback to the nineteenth century, that's what: a savage capitalist economy with no guarantees for workers and all power lodged in a voracious privileged elite of nearby entrepreneurs and distant corporations. Here we are, drowning in a retrograde, brutal, pitiless past, made all the more intolerable because the regime it had spawned commanded the most modern and up-to-date technology of surveillance and repression. And how, how, how to find a way out, how to get rid of Pinochet if we had been unable to foresee this apocalyptic miscarriage, how to persuade the trampled people of Chile to trust us again if we had led them to catastrophe. We had told them that the volcano would not erupt, the earthquake would not cleave us open, we had told them we could do what had never been done before in the history of humanity, a peaceful revolution, *compañeros, una revolución democrática y pacífica,* look at it, look at the blood in the streets and listen to the screams in the cellars, and look and listen to you, safe here in this cheap migrant hotel, alone with your ghosts and your guilt, remembering your friends who were not that fortunate, your friends who did not escape, your community that can no longer revive you.

I expended my misery as if I were a gutter, a sewer discharging into another sewer, lodged on that toilet seat a few inches above waters meant to carry away the shit and the piss and not to be listening to my fetid sorrow in the night, my lamentation for all the dead and dying and—

I stopped.

There was no time for this cemetery of words. Back home, right now, a car was stopping in front of a house and four men were getting out, a door was being broken into splinters, somebody young and radiant and blindfolded was being prodded down a flight of stairs, descending into a basement where he did not need to see the cot upon which they would lay him to know what was in store — there was too much to be done, I hadn't left Chile in order to wallow in self-pity. If this was all that I could spew out, all that I had inside, if I couldn't rescue one space of hope, one buried fire of redemption, if all I had to offer the world was more grief, then it was better to remain silent, that damn page had better remain blank.

I tore up my jeremiad, put the typewriter away in its case, clicked off the dismal light, and stealthily crept back into the bedroom. I waited there, standing in the steep darkness, listening to the miracle of Angélica and Rodrigo sleeping, praying for some divinity or devil to come and inhabit me and fill me with words, anything that would return me to the man I used to be, the country I used to call mine.

Nothing came.

I slipped into bed and watched the Parisian dawn, which had fascinated Monet and smiled on Picasso and dazzled Voltaire, invade my life, the cancerous dawn that would bring me no respite.

The next morning, the Algerian clerk behind the reception desk asked me if I had slept well, not bothered by clatters in the night, some maniac had apparently been typing away.

He knew it was me. Perhaps he had also been visited by specters from his North African sands, perhaps private labyrinths and casbahs frequented his daybreaks, and he was giving me a way out. I could have shrugged off responsibility. But somehow, after the nighttime ordeal with those toxic pages, after having told myself unwelcome truths, I couldn't stomach another falsehood.

"*C'était moi,*" I told him, in the wavering French that was not his native tongue or mine. I confessed that I was the culprit, and added that everybody could sleep away the night, certain that I wouldn't type one more word to disturb them.

That was a promise I could, regrettably, keep. It would be almost two years before I sat down again to write, before I found my true voice. I was still in Paris, and again I awoke just past midnight in an apartment on the Avenue du Petit Parc in Vincennes, but this time it was a nightmare I was escaping, two clowns had been torturing me to make me repeat horrible things about Allende, recite Pinochet's decrees, and I had surrendered my body and soul to them, rendered unto them every echo they demanded.

I slithered out of bed and went to the living/dining room area—the eighth residence we had occupied in Paris, this one loaned by a charitable friend. Almost three years after the coup and we were living off solidarity, we were living like beggars—and there, as ever, was my typewriter.

I wrote out the hallucination instigated by those clowns, let it tremble out into the world, then sat there, exhausted by the evil I was being forced to share, shivering with who I was, who we had become, what the world kept revealing about itself.

And as my heart slackened its beating, as I asked myself what comes next, after this what in hell comes next, I typed out something different: the voice of a blindfolded man describing how he counted the steps as they took him from his cell, if it was twenty they couldn't be taking him to the bathroom, *si son cuarenticinco ya no te pueden llevar a ejercicios,* if forty-five they can't be taking you out for exercise, *si pasaste los ochenta y empiezas a subir,* if you get past eighty and begin to stumble blindly, *a tropezones y ciego,* up a staircase, oh if you get past eighty there's only one place they can take you, there's only one place, there's only one place, now there's only one place left where they can take you.

Terrible, yes, and dark, but also dignified and extremely simple. It did not misrepresent what was about to happen, what had already happened so many times to that man, to other men, in those remote torture houses in Chile and across the world, it was not the subjugation those clowns had exacted in my nightmare, it was not the silence of the blank page. Later on, other poems — some even more despairing, some salvaging a residue of hope, a soldier who touches the arm of the man about to be executed, presses it gently, whispers please forgive me, *compañero*, and the prisoner's body fills with light, I tell you his body fills with light, and he almost doesn't hear the sound of the shots — other poems would come, then stories, finally a novel, a cascade of images that became unstoppable once something sanctified the night, opened me to the wonder of possible voices of redemption.

What had happened? What had made such a drastic journey out of silence even conceivable, what had changed?

For starters, I had.

I had needed time to accept that the language I had depended on during my whole adult life, the literature I had forged as my weapon against the rubble and absurdity of that life, was useless to me in my new circumstances. For that traumatized young man nothing could have been more natural than the inability to deal with the tragedy beleaguering him and his people. That sea of my sorrow could not be drained effortlessly because it was as real and vast as the death that had brushed by my body and stolen so many of my comrades and my hopes. I had never before had to figure out how to tell a story like this one, nothing I had written previously proved of value in this unforeseen predicament, no repetition of past formulas would or could bail me out. In penitence for my lost land, my lost self, I could not wish myself out of grief. Any crude formulation would have been fake, more of a betrayal than the stubbornness of a silence that wouldn't compromise.

What was happening to me, to us, was, quite literally, unspeakable.

Over those two and a half years following September 11, 1973, exile had, however, been working on me. Exile had savaged me, scoured me, skinned me with a knife until every smudge of my soul and every piece of my flesh was exposed and raw, and then it had poured salt on my wounds, and then had deprived me of the relief of telling anybody what I felt, had cloistered me in my regret, had not let me write one word unscarred or unchallenged by what had been inflicted upon us, not one false word, not one.

And by the time I had hit bottom, by the time my suffering had groomed me, something from the depths of Chile began to echo the extremity of that experience. History gave me company in my quest, a struggle by real people to parallel my own struggle to find a language that would not lie, my own bleak determination to vanquish silence had met the ferocious determination of the victims of the dictatorship not to disappear.

More than a metaphor, disappearance, *desaparición*.

Men and women were being kidnapped from homes and workplaces and streets, and the police and the government and the courts and the newspapers, anybody with a semblance of responsibility, turned the relatives away, denied any knowledge of the whereabouts of their loved ones.

There was a sick logic to disappearance, what made it such a prized form of repression. The authorities in Chile had their cake and could eat it too. Pinochet could kill my *compañeros* and at the same time avoid the public shame of engaging in mass murder; he could cleanse himself with official denials while being invested with the total power that comes from total terror. His regime's judicial accomplices could prevent writs of habeas corpus, because there was, to put it bluntly, no corpus. No body, dead or alive. No victim and, ergo, no crime.

No crime? The worst crime. Disappearance was an outrage against the chemistry and structure of life itself. The bodies of the missing were wrenched out of the normal progression of existence — the time of our species where conception leads to birth, and birth to childhood, and life to death, and from there to the resurrection in those who remain and remember. So it was not a passing coincidence that women became the central protagonists in the struggle to keep the missing alive, to demand justice. For the woman who carried the child, for the woman who had to carry that memory like a child, disappearance was the mother of all challenges. And because the dictatorship was depriving those missing persons of history itself, refusing to each body the only vocabulary left to it, the speech inscribed in that violence, the act of resistance by the relatives in Chile had to inevitably start in a speaking out, the telling of a story left incomplete.

Those women were using their imagination as a tool that defied extinction, the only place where burial, however uncertain, might be possible, setting to rest in the mind what could not be set to rest in the earth. Rebellion in the heart against the dictator's audacity, his resolve to be the only one to tell the story, was the first step towards other forms of rebellion, the primary miracle of affirming life in the midst of death. But for the voiceless writer who watched this battle for memory and justice from faraway Paris, those women performed a different sort of miracle. They seemed to be calling out, asking me to unlock myself and give solace to their loved ones. Those real voices of real women in Chile were creating a semblance of how a story can be told in the midst of stillness, a model of how a country can be kept alive in the midst of denial, a model of how a community can be rewoven in the midst of repression and bleakness, a model of how to hold out hope while acknowledging the despair.

They opened up a space of birth in the oppressive reality of Chile at the precise moment when I was ready to channel in some inexplicable way those other voices from the basements and the rivers and the mine shafts where the bodies had been hidden, the ocean into which those *compañeros* of mine had been thrown from helicopters. They came to me, victims and survivors, at the end of a night when I was finally able to listen and transcribe what something, someone — that man speaking to the prisoner as he mounted those steps, that prisoner up against the wall — was dictating to me so that at least the words would be there when we emerged from the haunted catastrophe, so we would not be entirely naked under the black sun.

A few months before my nightmare and the poem that had spilled out and into me as a response, I met Heinrich Böll in a café on the Rue Jacques Callot in the Quartier Latin to discuss how PEN's Emergency Fund could help writers persecuted in Chile and other Latin American lands. But the conversation soon turned to the topic of writing. Heinrich Böll! He had won the Nobel Prize, smuggled Solzhenitsyn's work out of the Soviet Union, authored novels that I had read with delight, and I thought, he will ask me what I am writing, how can I tell him that I am empty, that I have nothing to say, that I am just a party hack, that I don't deserve to be sipping this crème with him at this café. But he was a sage old man, considerate to a fault, and did not press me on what I might be doing. What he shared with me was the problem that German writers had faced after the Third Reich. "Hitler contaminated the language," he said. "We could no longer write the word *comrade*, the words *joy* and *exultation* and *brotherhood*. It was kidnapped, the language itself, by the Nazis. That was the task we could not avoid, that is what you must worry about most. Not allowing them to control the language with which you will tell the story of your times. This is something that needs to be done now, before you overthrow Pinochet. It cannot wait till tomorrow or it may be too late."

I nodded. What else could I do but nod and wait, nod and wait and see if the ceremonial dance of literature that had never abandoned me before would purify me one more time. I nodded and prayed that it was not already too late.

And when the words ultimately did come, when I started on this journey that I know now will end only with that other silence, the silence of my own death, I discovered something else about the writing that had visited me, a gift I had not expected.

My literature was not merely a territory where the dead could resuscitate, not just a funeral of words refused to the men and women who had been taken from their homes, who had been exiled into ashes, was not only a way of healing the self and the country held hostage by Pinochet as if it were a *Desaparecido*.

I was there. I was in Chile. I was mounting those steps. I was helping others to mount those steps with me. I was accompanying that man as he counted and recounted each step.

There was no other place left where I could go.

My imagination was taking me home.

Fragment from the Diary of My Return to Chile in 1990

JULY 24

First signs of trouble for our family.

When the Chilean people trounced Pinochet in the 1988 plebiscite, it seemed to prove that our citizens had not been polluted irreparably by their prolonged captivity. That epic victory against fear and an omnipotent dictatorship convinced Rodrigo, at the age of twenty-one, that his perpetual dream of returning to the land of his birth was now feasible, and he used a substantial college graduation gift from his grandparents to subsidize that desire to help rebuild Chile. So he arrived in Santiago eight months before we did, has had a chance to scout out the lay of the land. And, naturally, we were thrilled that the family would all be together once Angélica, Joaquín, and I also moved back here.

After a couple of months living with his aunt Ana María, Rodrigo has ended up renting, along with a couple of artists, a derelict studio in a seedy neighborhood downtown. The last we had heard, he was working as a personal assistant (though without pay) for an eminent Chilean actor, translating letters into English and a play from the French and running all sorts of errands, along with developing projects of his own in theater and video that would bear fruit, he thought, by the time we'd be settling in Chile.

A few hours ago, however, he came by — our first chance for a frank conversation — and announced that he's accepted an offer to work in a bilingual theater in San Diego and then will pursue a master's degree at the University of California. So by early 1991 he'll have left Chile.

What many of his generation would do if they had the means. Like myriad other young people here, he's disappointed in our newfound democracy, feels that it hasn't made that much of a difference in the lives of the majority. The very youngsters who fought the hardest against tyranny continue to be targeted as *anti-sociales,* are being beaten up by the same policemen as under the dictatorship. Most of *los hijos de Pinochet,* the children of Pinochet, as they mockingly describe themselves, are still unemployed, feel *marginados* from the transition. They look for solace in cheap drugs and alcohol, can't find adequate housing, and live in squalor, sensing that they have no future.

Just as discouraging is Rodrigo's personal experience. His boundless enthusiasm, curiosity, and energy, his determination to volunteer his time, his savings, his vision, have not been reciprocated. The actor he was working for has bled him dry and then reneged on a pledge to help Rodrigo put on a play, cast our son out as if he were a pair of old shoes, goodbye, *si te he visto no me acuerdo.* And that's typical of the attitude of everyone he's encountered with the slightest ounce of power: many promises and no follow-through, many hypocritical doors opened only to be slammed shut. You have to join some political party, he says, or be connected to the small mafia of the cultural elite in order to secure any real assistance.

"So your decision is final?" Angélica asked — and coughed that slight rasping cough of hers, the result of a tear-gassing during a protest in Santiago a few years back that left her throat scratchy and susceptible.

"If I don't leave, I'm going to die."

"You mean you can't breathe," I said much too quickly. "No space to breathe."

I saw Rodrigo hesitate, came to some sort of river in his mind and then crossed it. He said, "That, sure, but more than that. Killed, I'm going to get killed."

Some months ago, he told us, he and his two roommates had been arrested after impulsively rushing to the defense of a neighbor who was being harassed by the police because he was making out with his girlfriend on the street. A bit blatantly, Rodrigo admitted, but not breaking the law — loitering, the cops said, the same charge leveled at the three would-be rescuers. When Rodrigo had protested, a *carabinero* took out his pistol, put it to Rodrigo's head, and asked him if he was going to make this easy or difficult.

They were taken to the First Comisaría, which serves a district close to the presidential palace (and where, I note desolately, several of my *compañeros* from the time of Allende had been tortured to death), and spent the night there in the company of what Rodrigo called glue sniffers and petty thieves, the usual riffraff floating around the outskirts of Santiago. I wondered if the police realized that our son was somehow different. Or was he that different after all? Was I assuming that his class origins would protect him?

"I was in a holding cell," Rodrigo told us, "with maybe fifty other guys. And all the while I had been watching the lieutenant in charge, just keeping an eye on him — he was tall and of European stock, with an impeccably pressed uniform. I could tell he was pissed off at being there, surrounded by a beehive of nervous policemen, all of them with darker skin. Then he shouts our names and 'A la peni, you three, to the penitentiary.'

"And I thought, no way am I going to la peni—I won't get out of there alive. And I speak up: 'Wait, wait, wait,' like this, in a voice that was urgent and low, and I approach the lieutenant and he's up on an estrado, like judgment day, something out of Kafka, and what I said was only for his ears. I said, looking straight at him: 'You're white, I'm white. You have blue eyes, I have green eyes. You don't belong here, I don't belong here. What's the fine for loitering?'

"The lieutenant wavered for a second. He looked at my two companions—the neighbor and his girlfriend were long gone, who knows where they'd been dumped—and one of my friends was dressed up like a nineteenth-century ruffian, with a long cloak, as if he were the Count of Lautréamont, and the other looked scruffy but inoffensive, even if, in fact, he'd spent years in Nicaragua robbing banks for the Sandinistas—and then this officer looks back at me and tells me the fine is sixty dollars, and as luck would have it I'd cashed a check that very morning. I paid the fine and we were booked—disorderly conduct—and released. So when I say—"

"You're going to get killed—"

"I'm too wild, too free, and I don't know how to navigate this country, I don't know when to bite my tongue, when to speak. This time I managed to talk my way out, but next time . . ."

Next time his social class might not save him, might not help him stave off death. But I don't bring this up. It doesn't seem the right moment to indulge in sociological observations.

"No space to breathe, huh?"

"No space for someone like me."

And so he'll be leaving, repeating the path he took when he

left Chile as a small boy of six, but not because he is forced to follow his parents into exile; this time he leaves of his own free will, this time forced out by the country he kept so devotedly alive inside all through those pernicious years.

•

EXILE DESTROYS the children along with the parents.

After the coup, during my weeks of living clandestinely and then for months in the Argentine embassy, Angélica and my parents tried to shield Rodrigo from the insecurity and the horror, the lethal rumors swirling around him. Kids always end up knowing everything. Like so many other children of failed revolutions, our son was beset by nightmares, the same recurring vision that would replicate endlessly in the years ahead: some men wanted to kill him as he fled through a strange, unrecognizable city. And when he awoke, none of the white lies from his mother or grandparents could dispel what his antennae had picked up, that he was living no ordinary catastrophe. Your daddy's fine, he went off to the beach to write a novel, but better not to talk about him. Allende is fooling everyone, he's really alive and in hiding, but better to take his photograph down from your wall. School is closed, it's only for now, not true that your favorite teacher was picked up by the military, but better not mention him to your friends. Yes, we're going to Argentina. No, you can't take all your toys. Only one. Bring along the one you love most.

The oversized stuffed rabbit Rodrigo chose never left his side, was treated as if it were also a victim. One scene sticks in my mind: We are in our room at the hotel in Havana in early 1974, just arrived from Buenos Aires, and a Cuban *compañera,* Bea, with her thick glasses and pursed lips, concentrates on mending the rabbit, sewing back one of the eyes unhinged from so much trekking across Latin America. She sits on one of the beds and tries baby clothes on the rabbit, repairing it in the same way her colleagues at the Casa de las Américas have been caring for us after months of dread and humiliation, offering a transitory sanctuary before we left for the uncertainties of Europe.

By the time Rodrigo reached Paris, the rabbit had been joined by a collection of handsome miniature soldiers, a gift from my parents. In that dismal hotel on the Rue Blomet where we first disembarked, Rodrigo would spend hours playing on the floor — he was in charge of one cohort of soldiers while his stuffed pet commanded the other one. The rabbit was also receiving an education, as Rodrigo, precocious at seven, read a series of books to it featuring Mafalda, the Argentine whiz kid who was the most popular cartoon character in Latin America. Daniel Divinsky, my Buenos Aires editor, had given those comic books to him, "so you won't forget Argentina."

One day, as I was about to leave our hotel room to attend some smoke-infested solidarity meeting, the elderly French maid who cleaned up every other day ambled in and without a word got down on her knees and furiously began to scrub the floor with her fingers. Her target was some clay that Angélica had bought for Rodrigo and that he had spread all over the floor in a series of fortifications.

Angélica and I crouched down next to the irascible maid. In my flustered French, I attempted to dissuade her, she was as exploited as the women I was fighting for back at home, please, madame, we can take care of this, *s'il vous plaît, madame, nous sommes désolés, mais* . . .

She wouldn't listen.

"*Regards,*" she cried out in a broken voice, lifting the scruff of her fingers towards a crestfallen Rodrigo. "Look at my nails. Look at my French fingernails. Doing dirty work for foreigners. Look at what I have come to."

Astonished as I might be by the degree of her anger and chauvinism, I also felt that this caricature of Madame Defarge, with her worn-out knees and all the real pain of real floors scrubbed for paltry hotel wages, was speaking for me. Look at us. Look at our

Chilean hands. Look at my hands that a few nights ago typed out words of dereliction and now are incapable of writing even that. Look at what we have come to.

Fortunately, we were soon to leave that hotel. We had been invited to a Chile solidarity rally in Reggio Emilia, in the north of Italy, and we hoped, on our return to Paris, to move into an apartment now occupied by Claudio Iturra, a friend from way back, a *compañero* from the Chilean Communist Party, famous for having penned the lyrics of "Venceremos," the hymn of the Allende revolution. "Solidarity rental rates," Claudio said to me, "courtesy of French comrades." As it made no sense to haul our scant earthly possessions onto the train, we stowed everything in a closet in Claudio's apartment, soon to be ours for a pittance. Rodrigo couldn't conceal his anxiety as he watched his prizes deposited up on a shelf.

At least he'd kept his rabbit, one stable friend in a hostile universe.

Not for long.

We lost that rabbit in Rome.

Some minutes after we checked out of our hotel on the Via del Corso, on our way to take the train back to Paris, Rodrigo realized he had left his favorite stuffed animal behind. We ordered the taxi to turn around, galloped up the stairs. The room had already been vacuumed, the maids had seen nothing. We raided the garbage, we combed each floor, we left no corner unexplored, we lost our train, we came back the next day, nothing, nothing, nothing.

Rodrigo remembers the sequence differently. He says we weren't leaving for Paris that day but going to a friend's house in Rome, and that we refused to return to the hotel straightaway, promised to look for the rabbit the next day, wanted to teach him a lesson, to be more careful with his belongings. Maybe my version was constructed to alleviate my guilt and his in order to emphasize his abandonment. If I have to choose, Rodrigo's alternative makes more sense, brutally illustrates how frantic our days were back then, how inured we had become to pain.

Bad news in Paris. Claudio Iturra explained that he had been forced to leave the apartment because of threats he had received over the telephone, possibly from Pinochet's secret police. The Communist hierarchs of his party had deemed the place unsafe, to be abandoned straightaway. We were upset, of course — where would we find solidarity rates like these? Our pilgrimage across Paris in search of a residence was going to start in earnest. The only question to an embarrassed Claudio: When could we pick up our stuff?

Juan, a member of the Communist Party's security apparatus, would assess when the pressing danger had abated and set up a visit. By the time he called us, Rodrigo had been packed off to a summer camp organized by the French trade unions for Chilean kids, a month in the Alps, all expenses paid — an expedition arranged by none other than Claudio Iturra.

Juan — if that was his real name — was waiting for us in the apartment, his wife hovering behind him. What was she doing in a place that risky? As if to stave off questions, they hastily led us to the closet. Only the two suitcases with our clothes were there.

"*Y las otras cosas?*" Angélica asked. What about the other things?

"What other things?"

"There were books and toy soldiers."

"Just junk," Juan said. "Thrown out when we vacated the apartment in a hurry." And when we remonstrated, he added something about kids in Chile being raided by the police every night, why the big fat fuss?

It was an argument we were to encounter over and over again, the typical — and idiotically effective — moral blackmail of exile: you people are worrying about something as insignificant as . . . (fill in the blank). You're complaining about this when Chile is under the boot? You people have money for toys, and the children back home are lucky to eat stray alley cats!

We knew that Juan was lying about Rodrigo's belongings. No Latin American in his right mind would have discarded those Mafalda books, outrageously popular with adults. As for our boy's lovely soldiers, they were probably in the hands of some other child, maybe the son of this very man with his self-satisfied proprietary air, contentedly installed in this presumably perilous apartment. When we peeked into the kitchen, where his wife was preparing a stew, we noticed a toy duck lying on the floor. In the relentless scramble for resources, his party—which was not ours!—had decided to take care of its own militants first. Claudio Iturra had been pressured by some commissar to pass his plum lodgings on to a comrade of the same Communist persuasion.

"*Hijos de puta!*" Angélica muttered as the door closed behind us, loud enough, I both hoped and feared, for those two inside to have heard. They were more malevolent than that bedraggled, half-mad cleaning woman in the hotel. At least that hag was no hypocrite, she did not try to edulcorate her disdain with words about brotherhood and revolution while flapping her cracked French fingernails at us. She hadn't stolen from a boy lost in exile, wasn't evicting fellow revolutionaries from their lodgings, was more of a comrade than this Juan.

And yet, as soon as we descended to the Paris streets with our suitcases and our self-reproach, the nasty, contradictory truth of our condition assailed us. We were far from home, the war against tyranny had not ceased, and that man and woman, no matter how contemptible, would be part of the struggle, were still my *compañeros*. The same people who had purloined my child's soldiers would have worked their heads off to free me from jail if the real soldiers back home had arrested me, were working right now to help liberate us from the plague of Pinochet. And Claudio, who had cavalierly handed his associates the apartment pledged to us, had just spent weeks selflessly organizing a vacation for Chilean kids.

To stay angry at people I would be working with for years to come made no sense, especially when there was an easier target. I had been the one, after all, who believed that the purity and enthusiasm of the days and nights of an ascendant revolution could last into the sad spiral of defeat and beyond, that the same laws of fraternity would prevail, that I would find angels in my path now that I had been ejected from paradise. Though it had not been that naïve a presumption, not that wrong of me to expect that our fellow exiles would be, not only *compañeros*, those who break bread, share their *pan*, with you, but *com-padres*, sharing their family and their future, godfathers to each other's children, caring for them as if they were their own, acting as if those boys and girls displaced from their homeland were our collective responsibility. The usurpers of that apartment should have been my home away from home, should have shown a sliver of compassion, offered an anticipation today of the sort of society we wanted to build tomorrow. Should have made it easier for our children to forgive us for what we were making them go through.

It was the first time in our banishment that we came face to face with what it meant to have lost the community we had once belonged to, the community we would need to keep alive if we were ever to bathe again in the same radiant river. I had watched my *pueblo* come out of their own exile, make theirs that alien land they had mined and built and fed and fought for in endless wars and loved in endless beds, briefly take possession in the Allende years of streets that had never been named for their mothers and forefathers, occupy, for one prolonged transitory moment, the multiple, spilling spotlight of history and hope. Those dirt-poor people who had only their deaths-in-life to live and their half-lives to die had taught me that nothing on this earth is eternally fixed, nothing has to be the way you found it at birth, that we have something to say in all of this before we disappear into the winds of time. And now they were expelled again, foreigners again in their own land—and what was that doing to us, to all of us, that expulsion? What other betrayals lay down the road, would befall us before we all went home?

In the end, Rodrigo took the news that his books and toys had vanished with more equanimity than his parents did. Maybe he had figured out by then that there was a price to pay for having survived and that he might as well start servicing the first installments of the debt as soon as possible.

Whatever the reasons for his wisdom, it was a miracle that Rodrigo did not let that act of duplicity and spitefulness demolish his trust in others, his wish, perpetually renewed, to live permanently in Chile, claim the country of his birth as the country of his choice.

Alarming, then, that it should have been Chile's transition to democracy in 1990 that shattered Rodrigo's commitment to his country, which had not yielded all through the years to the slings and outrages of exile. I could not know that he was presaging what would happen to me six months later, that this time it was the father who would follow the son into a wandering world.

But perhaps this tone of sorrow and nostalgia is not appropriate in 2010, is a residue of a sadness no longer part of my life. We may not be in Chile together today, but the entire family lives in one place, in Durham, North Carolina, less than ten blocks from one another, and there are two granddaughters who illuminate our days, and Rodrigo and I have become collaborators in film and theater and intellectual conspiracies, not knowing who is the mentor and who is the protégé, who teaches and who learns. Can I really ask for more?

Back then, however, I was afraid we were going to lose him. And all I could do to comfort my grieving heart was to remind myself: this son of mine is now an adult and must seek out his life where he feels he can give most of himself. At least that's what

I recall: comforting myself and Angélica, and of course Joaquín, so visibly affected by the news that his brother would not stay in Chile, wouldn't be close by as he tackled his own problems of settling in. Not to worry, I said to Joaquín, perhaps more to myself than to him, Rodrigo may change his mind, could well decide to remain near to us after all. And only to myself, as if whispering hidden words into my diary: maybe the country will understand that it shouldn't be wasting its children like this, maybe a few years from now, I murmured to myself in 1990, he'll be back and our family will once again be reunited under the same sky he left so long ago, through no fault or decision of his own, I swore that we would work this out, how could I accept that he had succumbed to the curse of dislocation that the men of the Dorfman tribe seem unable to escape, that the women of the family suffer along with us?

Fragment from the Diary of My Return to Chile in 1990

AUGUST 4

A blast of wind makes the glass doors of my small study rattle, this Southern Hemisphere winter chilling me to the bone. I wrap the long poncho more closely around my legs. I hope I'm not coming down with a cold.

This damn house. This damn wonderful house.

We bought it in 1986, as soon as a permanent return to Chile seemed viable, a plan facilitated by an agreement I reached with Duke University. My salary for teaching four months each year in Durham might be meager, but it would be enough to live austerely the rest of the time in Santiago. We didn't mind that this charming small house *en la calle Zapiola*, acquired in the midst of dictatorship, was flimsily constructed. Paramount was security—to know all your neighbors, each and every one an opponent of Pinochet, so if the secret police came to cart you away in the night, there would be witnesses, so the army couldn't unload a weapons cache in your backyard and accuse you of terrorism. When we first signed the sale papers for this perfect place to settle in for good, we didn't pay attention to the vines creeping through the brick walls, the unfinished, hungry look of the fixtures, part of a plan by the original architect to "respect Nature." Now that we have to live here, have spent these weeks patching the house up, its flaws seem less quaintly enchanting.

If it were only the cracks through which the cold air whistles. The bricks leak, the faucets drip, the keys don't fit, the fuses blow when Angélica plugs in the clothes dryer, and the voltage . . . don't get me started on the voltage and the firetrap installations, how the doorbell tingles you with a slight electric shock. Taking a shower is an adventure: you can't guess when the capricious gas burners are going to peter out, forcing you to relight the califont heater with a soggy match in your trickling fingers and then shiver back into the shower, only to hear the gas cease its roaring, and the plumber, the *gásfiter*, promises to hurry over *hoy mismo* to rejigger the shabby work he's already botched a few times, and we wait for hours and he doesn't even call to excuse himself. This is the country of *mañana* and *perdona* and so sorry. And this study has no heat: when I'm done writing these solitary musings in my journal, I'll have to cover the computer and the printer with blankets bought the day after we arrived, one of many trips, because the deed to Zapiola has been inaccurately registered, a legal snarl that four different offices on three different mornings couldn't untangle, and no bank will open an account for us because I don't have a job in Santiago—what do you mean you teach at Duke University in the United States, what about here?—and all these errands are further complicated by periodic breakdowns of the old Peugeot we bought several years ago and which has also turned into a clunker.

In fact everything here seems to be a clunker, especially in contrast to our smooth and functional life in the United States. We're spoiled by our life abroad, that's the truth. We've grown accustomed to telephones working and handymen arriving on time and smiling tellers who welcome you to the bank, and the idea, absurd to Chilean merchants, that the customer is always right. But hey, if I wanted to be cozy and sheltered, I should have stayed back there in Durham, where a house furnished from top to bottom awaits us and a well-oiled car and a postal service that delivers mail without the carrier demanding a tip for each item.

A few days ago we enrolled Joaquín in Nido de Aguilas, an American-inspired school we hope will facilitate his transition — not plunge him directly into an exclusively Spanish-speaking environment — and a recently arrived Canadian woman told us while she waited to register her own son that she had placed her affairs in the hands of two resourceful local ladies who, for a fee, had negotiated a complete installation package, made sure everything was working to perfection, just like back home. Angélica looked at the woman with something bordering on envy — to alight in a country that way, *qué maravilla!* But even if we could afford such extravagance, I wouldn't want to do things that woman's way, wouldn't want to live like her, inhabit an enclave of modernity that shuts Chile and its foibles out. Damn it, for us this *is* back home! I'll probably spend years squeezing the errors and blunders of underdevelopment out of our Third World residence, carping about the irresponsibility and incompetence of our Third World country. I'll be incensed at every wasted hour, but another part of me — not the Ariel who tomorrow must get the kerosene siphon for our only heater fixed for the third time — another part of me is grateful for this recalcitrant welcome that translates us, stuttering, into the code of the country.

This is the Chile that destroys time as predictable, repetitive, humdrum, *rutinario,* a Chile that breaks the clocks and the doorbells and the microwaves into shreds so you are forced to imagine life in a different way. Just yesterday I wrote to my parents in Buenos Aires with an ode to the potholes that disgrace our streets and devastate our cars — it is Chile, I told them insanely, Chile that is warning us to stop speeding towards an unsurprising, conventional, United States kind of existence, a rebellious Chile that is conspiring under the cement of newfangled consumerism and high-rises and supermarkets to cleave open the floodgates of reality, to pothole you and shake you and fragment your every certainty.

That's how I defined Chile from exile: as a catalyst destined to explode into pieces the life we've led, sow a foundational chaos. And that's why it was easy to postpone any major self-scrutiny as we wandered — it was more than enough to have to change houses and countries and vocabularies and grocers and bus routes and address books; that was sufficient, thanks. The postponement made easier by the expectation that someday, when the physical journey was over, when Ithaca was no longer a mere mirage on the horizon but a palace where we'd sit by the old fires of yesteryear, there would be time to cope with the blemishes and limitations. Chile was supposed to be a transformative leap into the maelstrom of experience that would reorder my priorities, a new Ariel for a new era.

I love it here, love being back with such utter abandon that I turn even the worst problems into a source of perverse enjoyment. The other day a frustrated Angélica came home with a faulty new blender that the store had refused to replace or fix — twice! *Basta,* she said. You go do it, Ariel. You're tall, you're blond, you're not afraid of speaking up. So I stuffed blender and body into a rickety bus. In Santiago, you never know how long it will take you to get anywhere. Drivers dash frantically for the next stop, trying to beat their rivals — or if they're too far behind to win the race, they'll trundle along at a turtle's pace, hoping clients will accrue, which means that passengers spend long periods waiting for transportation under a gray drizzle. This erratic rhythm is determined by urchins who inform the driver of the whereabouts of possible competitors, whether to slow down or accelerate. This drives (*sic*) everybody here berserk, but I tend to be amused — perhaps because I have time on my hands, I can let the country seep into me, send me secret messages. On this occasion, I descended into the coughing *centro* in a relatively benign mood. Which quickly soured as one salesman after the other gave me the runaround, until I extracted from one benevolent soul the address of company headquarters — just across the street — and crashed my way through an army of secretaries. Sir, you can't go in there. Oh no? Just watch me! And the boss, cowering behind his desk, managed a timid, Sir, don't raise your voice like that. Raise my voice? I haven't even started to raise my voice. The one thing Chileans fear above all else is a scandal, to look ridiculous, so the browbeaten exec commanded that the blender be replaced.

But my triumph left a bitter aftertaste: in order to get things to work effectively, I'm in danger of becoming a *prepotente,* an arrogant prick, like the people in power in Chile all these years who feel such a sense of privilege that they live complaining and bullying the lesser beings around them. I've riled my tranquility, all for the sake of a stupid blender. I just can't blend in, I tell Angélica, trying to tease out a pun from all this, but she doesn't laugh, looks at me with zero commiseration. Aren't you the one who thinks everything is fine and dandy? Because she doesn't, she doesn't think this is going to work out, *mi amor,* that's what she said to me last night, standing in the passionate middle of the living room, I really don't.

She came here full of enthusiasm, ready to give her all to the new Chile, more than ready to work again in the *poblaciones,* to help disadvantaged women just as she did during the

Allende era and all through our years in Washington, D.C. I
don't expect to be paid, Ariel, all I want is to contribute. It has
taken her less than two weeks to realize what it took Rodrigo
eight months to discover, what according to her I am too stubborn to admit, the truth she won't back down from: this transition to a restricted democracy has no place for someone
like her and definitely won't have a place for someone like me.
She's willing to stick it out for now, fire up her engines all over
again, wait until I come to my senses. But it's also clear that, after all those false starts and new beginnings of our odyssey, my
Angélica keeping our stranded family sane in every transitory
home of our transitory lives, a new dentist and a new market
to shop for the least expensive tomatoes and a new map to decipher and new regulations to memorize and a new school for
Rodrigo and patching up his clothes far into the night, the elbows and the knees of his clothes to stretch our budget, all
those cities, the endless stream of bad news and aborted expectations, always on the move, plane after plane after plane
and a van from Paris to Amsterdam and a ship to cross the
Atlantic to Baltimore, packing up each time, suitcases like
closets, nothing ever put away entirely because tomorrow
Pinochet will be gone and we'll be heading home, and even
once we were allowed back in 1983, still seven years of more
planes, back and forth, closets opened and closed, suitcases
closed and opened, back and forth, it's all too clear that after
nomadic years of improvisation, she's the one who, as usual,
will bear the brunt of this dislocation, this final one, what I
hope will be the final one, the end of our journey home.

Yes, it is final.

My evidence: the new mattress we bought a few days ago
here in Santiago.

Those who have never suffered the iniquities of exile cannot possibly understand the significance, the gravitas and
gravity, of a mattress. Not because you'll spend one-third of

your banishment sleeping on one and dreaming of going back, scared that you might awaken in the night sweating with the sense that you may have started to dream in a language other than your native tongue. And not because that is where you make love, find in your soul mate the one temple that still stands, that is not defiled.

Something far more prosaic: mattresses are large, cumbersome, expensive, difficult to transport.

During the first nine years of exile we did not buy even one, not one. We perpetually put off that purchase, experts in thrift-store furniture and loans from other transients, everything in our temporary abodes ready to be discarded the instant the radio brought the flash that General Pinochet had been overthrown. Even after we got stranded in the States in 1980 and settled in for the long haul, we still slept on a borrowed mattress for an extra year. Astonishingly, it had been made in Chile — inherited from a long line of compatriots, imported to D.C. years earlier by a long-gone diplomat. Its generous and dilapidated hulk was finally replaced by a new, heavy-duty, extra-firm, dream-worthy wonder.

A Sealy that awaits us in Durham once we go back for the spring semester, so I can teach at Duke and pay for next year's return to Chile. Whenever people have asked me what I'll be doing now that I'm back, my answer has been: nothing. I will do nothing for these initial six months and I won't look for work in subsequent years either. There is relief and bafflement in the questioner's eyes when I insist I don't want a job, prefer not to be seen as a competitor fighting to occupy space that should be reserved for those who have spent their days in the shadows of the struggle, that they should have priority — no job, not in the university, not in the government, not in the vast civil network of NGOs.

I'll make my living by teaching part time in America until, a few years down the road, say by 1994, I resign from my blessed Duke position and try to live off my writing. Soon enough I'll be ready to forsake my vow of silence, and words will come pouring out of me like a maniac. I have a contract with Pantheon for a book, and my latest novel, *Mascara*, about a faceless man who steals people's souls with his clandestine camera, is being translated into several languages and my theatrical adaptation of *Widows*, coauthored with a talented young playwright called Tony Kushner, is slated to open in Los Angeles next year. So a day will come when we can sell that Sealy mattress that grows old in Durham or give it away to some other refugee, we'll gradually transfer our belongings from there to here — too complicated to think how that'll work itself out, too much anguish to wonder about the library I've accumulated abroad, what to do with it, how it will feel when we cut ties for good with the United States, something to think about tomorrow, the day after tomorrow.

For now, this is the only roof that matters, this roof above our head that separates us from the stars burning in the Southern sky and yet, at the same time, joins us to them.

A roof above my head.

What every mortal since time before time prays for.

The primary implication of such a longing was understood by the Romans when they became the first to legislate on this matter, as on so many others. Those laws determining the conditions of banishment were remorseless and prescient: the expatriate was to be denied not only *aqua* (water) and *ignis* (fire) but something else, entirely social and architectural, something that needed many hands to build it: *tecte*. A roof. So the offender would be at the mercy, as the exiled Virgil put it, of "another sun," threatened by the lightning that can strike us from heaven and the rains that drench us as we roam. No *tecte*, meaning that we cannot be sustained by the rituals our forebears created in order to capture and domesticate water and fire, no hearth to sit by, no jug to drink from, no pro-tect-ion.

Maybe that's the lesson I had to learn during my banishment. Returned to the nomadic existence that had once marked us as the species of Cain, weary like our ancestors of the dizzy, starless nights, desperate like them for a refuge against the teeth of predators, I hankered, in Paris and Amsterdam and Washington, for the reaping and roof of Abel, replicating in my own strange way the journey of mankind from the trees and the savannah to the cities, as men and women had done for millennia, always in search of some prayerful stability, a permanent pillow for my head.

But enough meditation for this cold August day, enough memories of the faraway mattress, made and bought in the United States of America, that incited my dreams of faraway Chile. Last night, in nearby, oh-so-close Santiago, our bed began to cave in because the carpenter miscalculated the length of the planks to uphold our new Chilean mattress. I will disengage it this afternoon and lubricate the joints with soap so the wood won't squeak.

I wouldn't want to wake up in the middle of the night and realize that I have been dreaming about the United States and a faucet that does not leak and a blender that does not break down, dreaming of a bed that will not collapse before dawn arrives.

But no matter where I awaken, where I ultimately end up, Angélica will always be by my side.

•

Angélica, Angélica.

How can it be that those we love most, that the woman deepest in my heart, had to pay the price for my follies, that's the real question. Is that where love leads us?

Angélica had blossomed during the revolutionary years of Allende. At the time of the coup, she was on the verge of finishing a postgraduate degree in education. Though the textile factory where she'd been counseling workers on how to continue their studies had been bombed and some of her protégés killed or jailed, Angélica didn't seem to be in immediate danger. She left Chile because of me, lost her country because of me.

Not the only thing she lost. Once abroad, she soon realized that her hard-earned female self-sufficiency was under siege. Like most women in exile, she was being yanked back to a dependent status, spousehood as the primary source of her identity. That I had been ordered to leave the country to carry out a major mission — to organize support abroad for the cultural resistance, to create a network of foreign journalists who would inform the world about our struggle — made me, by definition, valuable, and my wife, by that same definition, secondary and derivative.

I was not a traditional macho. More than any other Chilean male in our social circle and beyond it, during the first years of our marriage, before exile tore us away from a normal life, I shared the housework and quite a bit of the child-rearing, to the point that many friends who had never washed a dish or run the vacuum cleaner or changed a diaper would often wrinkle their noses and deride me as the *mamá* of Rodrigo, although nothing I did compared to the energy and affection that Angélica dedicated to the myriad details of our domestic life. Those "time-honored" tasks conventionally assigned to the female sex — cooking, caring for children, finding the cheapest way to clothe the family — became even more vital in exile. And more burdensome because they had to be accomplished without the safety net of relatives and friends who watched out for each other, the sort of bonds that don't seem indispensable until, when most needed, they evaporate from one day to the next. And Angélica also lacked the helping hand of another kind of woman, a woman who helps because of indigence rather than love, the maid found in all moderate-income residences in countries such as Chile and who allows the lady of the house her freedom of movement and the man of the house his freedom of conscience — because it isn't his wife slaving away.

After our marriage in 1966, we had rejected that demeaning exploitation, wanted to walk without clothes through our house and without guilt through our life. But the lack of aid at home segregates you from your milieu, becomes problematic when four or five friends drop in for a bite at midday or midafternoon or midnight, and this in a world with no frozen food, no labor-saving devices, no babysitters, no fast-food joints, a society that doesn't expect you to show up at dances and dinners with a kid in tow. And so, we had grudgingly hired a maid.

I can remember back then secretly pining for a day when we would again live without a stranger genuflecting in our midst. But when that dream came true in exile — not out of any devotion to equality but because we could scarcely pay for our own sustenance, let alone somebody else's salary — it was Angélica, of course, who ended up tied to the domestic world she had unshackled herself from, Angélica who found herself ever more subordinate. Workers had lost their trade unions, citizens had lost their freedom of assembly, peasants had been dislodged from their farms, prisoners had lost their right to habeas corpus, their right of representation, their right to a trial or even to an accusation — everybody had lost something, and as usual the women lost more than the men. With things falling apart, how to find the leisure to explore a less dominant relationship between the sexes, how to avoid devolving into a more traditional connection between man and woman? The distance in power and knowledge between male and female, which the Allende revolution had successfully been reducing, was reestablishing itself with a vengeance as our country retreated into a Victorian Age. As for me, no restrictions on Ariel. He had to be free to help dethrone Pinochet so our country could get its life back, so we could get our life back, our maid back. He had to be free to tell the story. And his wife would simply have to postpone her own goals, her own future, her own story.

We discussed the situation one painful, serene afternoon in Buenos Aires in January of 1974, on the hill that overlooks the

Torre Inglesa and the Retiro train station. Angélica had been thinking of leaving Argentina to return to Chile. Our marriage was in trouble, Rodrigo was a mess, she could already forecast years of tribulation. But what to do, if she loved me, if she was willing to do anything and everything to keep her family together, to avoid replicating the sort of divorce that had embittered her parents? What to do if her first name and her middle name and her last name have always been loyalty? And what would be my fate if she left me, not only taking away my life's companion and our son, but also, for this perennially uprooted man so recently displaced from his adopted nation, taking my one sure link to its lullabies, legends, jokes, accent, slang, ancestors, gestures, folk dances, meat pies, my wife as my own intimate territory, my substitute homeland, the essence of Chile? I blush today at how conventional this sounds, the male who digs and mines the female earth, the female as custodian of heart and hearth for the wandering male hunter and warrior, but that is how I felt then and still often feel. I'm great at exaggerating, but no exaggeration can convey how grateful I am that she decided to remain by my side.

But not grateful enough to say no to the party when those supposedly superior minds determined, one afternoon in a hotel in Havana in February 1974, that Paris was to be my destination. By then Angélica and I had heard reports about that city, wonderful if you were prosperous but the worst place to survive, unforgiving and haughty, if you happened to be poor and depressed. Angélica had said before the meeting where our fate would be sealed, *Por favor, Ariel, cualquier lugar salvo París,* anything but Paris.

"Paris is where you're needed." The first thing Enrique Correa said to me in that room in Havana. "That's where the Resistance will set up its cultural and press campaigns and where you'll be most productive."

How could I argue with Correa? He had just come from four months in the Chilean underground, was already a legend as a man who would soon sneak back into our land under a false identity, how to say to him, *No, no quiero,* I don't want to, I can't, I won't? Had I not tried to distinguish myself from other Chilean militants who had first sought asylum in foreign lands? They had acted out of necessity, because they had been denied a passport by Pinochet, but also because they were mature enough to know that it made sense to receive health care, job training, language lessons, special schools for their kids, and a supplement to rent a house, even money for furniture, all of it paid for by generous host governments.

Not the path I gloriously took.

And paid for that pride with years of suffering, years of wandering, not just for me but for my family.

So many houses, in Paris alone. Sleeping in the apartments of friends, six or seven I think the count was, before Angélica arrived with Rodrigo from Havana in May 1974, and we wound up in that foul hotel on the Rue Blomet, followed by that trip to Italy and then back to Paris, where the Communists denied us the promised lodgings, the initiation of an endless odyssey, a month in a desultory flat near the Rue Monge, rented from an enigmatic Vietnamese lady called Madame Nguyen, until free digs on the Rue des Canettes were loaned to us by a Mexican poet and his French-Tunisian wife, who were heading to Mexico for a few months. The owner, a generous friend of theirs from their bygone Maoist days, who went by her childhood name of Bidule and didn't live at the Canettes apartment, didn't charge a cent to any lodgers, didn't even let us pay the gas bill.

The flat consisted of one narrow, long room in an old sixteenth-century building, with a shower overlapping the kitchen sink. The toilet was out on the fourth-floor landing, distinguished by a swing lowered from its hook for defecation purposes. Rodrigo amused himself for hours with the Tarzan toilet. That, and walking down the street to the nearby Église de Saint Sulpice to play with candles meant to intercede for the dead and dying.

And then, starved for fresh air, and bolstered by a small one-year grant I had managed to squeeze out of a foundation established by the German Social Democrats, the Friedrich-Ebert-Stiftung, we rented an inexpensive and petite cottage in the suburb of Palaiseau. When our cash ran out, we moved to munificent Bidule's second apartment in Vincennes, loaned by her for free while she went off to live in Havana, of all places, with a graphic artist she'd fallen in love with during a solidarity visit.

As I spent most of the day working gratis for the Resistance, we barely got by on scant royalties from a few of my books and what Angélica could scrape from instructing French kids in English at a nearby school, plus extra francs gleaned from her babysitting. Our budget was so tight that I habitually jumped the subway turnstiles, sure that I'd never be caught. If I glimpsed inspectors checking *les tickets du Métro,* I'd wait until an African passenger came along, or someone of obvious Arabic ancestry, and he would always be checked (as Angélica, with her Mediterranean looks, invariably was), and I'd slip by, shielded as much by the conspicuous brownness of others as by my tall European build and Jewish nose. They couldn't imagine, those inspectors, how destitute and unstable my existence, my income probably lower than the migrants they were stopping. Not that we were in danger of sleeping, like *clochards, sous les ponts de Paris.* But penurious enough to qualify Rodrigo for free lunch at school, while his parents often ate at the dreary Sorbonne cafeterias, where I felt out of place, the oldest person in those halls bustling with students young enough to have been taught by me, like the ones in my classes not so long ago in faraway Chile.

Of course I hated this life, hated that idiotic Ariel for willingly immersing the family in that state of deprivation. I hated being invited out to dinner by French or Latin American friends and demurring because I couldn't reciprocate or help pay the bill. I hated having to recur to an amiable French doctor who did not charge us a penny one night when Rodrigo ran a fever and had trouble breathing, that doctor who gave us free medicine, I hated the unmitigated pity I saw in her eyes.

But above all, I hated Paris.

I hated Paris when it drizzled and I hated Paris when it sizzled, I hated April in Paris when the chestnuts blossomed and I hated October in Paris when *les feuilles mortes* fell and I hated Maurice Chevalier who had tap-danced along its boulevards and I hated the Hunchback of Notre-Dame who had been rejected by everyone setting eyes on his deformed, neglected, unwashed body and I hated the tourists at the Tour Eiffel and the view from the Tour Eiffel and the Tour Eiffel itself, and I hated even more the Champs Élysées where Jean Seberg had sold the *Herald Tribune* and betrayed her Belmondo lover, and I hated the offbeat markets of Montmartre and the wondrous façades behind which Rimbaud and Ronsard and Verlaine had cast their poems into the world like drunken ships and I hated the pavement where Molière had stubbed his toe and turned his lurch into laughter and I hated the Impressionists who had made the haze of those skies immortal and I even hated the back alleys where Henry Miller had fucked his brains out, I hated the *boulanger* who told me he would not serve me that *pâtisserie* until I had learned how to pronounce the word properly and I hated the supercilious way in which the inhabitants of the sacred Cinquième Arrondissement repeated back to me the French I stumbled through and I hated the shop windows full of the most scrumptious *jambon du pays* and *escargots,* I hated everything in Paris and about Paris and around Paris, every hour of my two and a half years there, every minute, every second, my oh my, did I hate Paris.

That's what I proclaimed at the time and for many years afterwards, even if it was not entirely true, even if I was inflating my distaste then and also now for poetic effect, even if Paris is today, as I write these words in 2010, our favorite city on the planet. I probably would have been equally miserable anywhere else, but as

I happened to spend the first scorched-earth years of my banishment in Paris — and it was my own fault, I was the one who had not been able to rebuff the party's orders! — I ended up blaming it for my troubles, as if the City of Light were really responsible for my insolvency, my multiple humiliations.

I hated my life there, yes, but simultaneously embraced it as a badge of honor, masochistic proof that I was roughing it, a true soldier of the revolution.

"We have to sell our house in Chile," Angélica said to me one day in Vincennes, not daring yet to demand that we abscond from this city where we were so unhappy. "We have to sell it and use the money to tide us over while we figure out what to do next."

Even if it was an unassuming bungalow on Vaticano Street in Santiago, we had hesitated before accepting my parents' help to buy it in 1970, four years after our marriage. Our hesitation derived from the conviction that to possess a residence would make us too bourgeois, probably the worst insult that could be launched at any young person in the late sixties, a betrayal of both our revolutionary and our pseudo-hippie credentials. We didn't want to own up to being privileged, the fact that my progenitors could give us ten thousand dollars for the purchase in a country where most of those marching with us in the liberation movement had to survive on a couple of dollars a day. I can remember telling Angélica how scared I was to be tied down to one place, what that might do to our dreams of absolute freedom, our engagement with poverty as an option, our desire to breach the distance from what we called "the masses." I was infused with the idea that property corrupted, made one into a defender of the status quo. This, at a time when we had no telephone and no car and no particular amenities, were barely getting by on my university salary, supplemented by private classes for rich kids in need of tutoring, along with the minuscule income from my critical reviews of literature in *Ercilla* magazine. Eventually, of course, we fell in love with that house of ours and, once we were far from its delights, revered it as a sort of sanctuary awaiting our return, the nightmare of exile no more than a parenthesis, we'd open our eyes one day and be back.

I had withstood selling the house on Vaticano up until then because of my library. Those books, full of scribbled notes in the margins, had been my one luxury in Chile, companions of my intellectual voyages, my best friends in the world. I had poured any disposable income into that library, augmenting it with hundreds of volumes my doting parents acquired for me. It was a collection that overflowed in every impossible direction, piling up even in the bathroom and the kitchen, until a few months before the coup we'd paid a carpenter to turn our one-car garage into additional library space. Angélica and I joked that Pinochet's military takeover had come just in time to avert a marital crisis generated by the ever-encroaching books.

It was a daily comfort, in the midst of our dispossession in exile, to imagine that cosmic *biblioteca* back home, gathering nothing more lethal than dust. That was my true self, my better self, that was the life of reading and writing I aspired to, the space where I had been at my most creative, penning a prize-winning novel, many short stories, innumerable articles and poems and analyses, in spite of my own doubts as to whether literature had any place at all in a revolution where reality itself was more challenging than my wildest imaginings. To pack the books away would have been to admit our wandering as everlasting. Even buying a book was proof that we intended to stay away long enough to begin a new library.

"We need a French-Spanish dictionary," Angélica would say to me as we roamed the outdoor bookstalls along the Rue de Sébastopol. "Look, here's a used one, not in bad shape."

"I have six in Chile" came my unwavering answer, embellishing my predicament, as always, but not inflating the number by that much — four maybe, four in Chile that I had rarely opened, and none in France where we would have needed to consult a dictionary several times a day. How do you say in the language of Camus and Balzac that we were fucked?

Nous sommes foutus, that's how, that's what we were, one phrase we had learned, heard too many times over.

"We've got to sell the house, Ariel."

So it was with regret and even torment that I agreed to Angélica's suggestion. But a few days later, I was saved from following through by a visit from Juan Enrique Vega, a party undersecretary. He plopped himself down in our living room and, after sipping some tea, announced, almost in a whisper:

"This is confidential, and it can't go beyond this room."

We waited expectantly.

"It's about your house, the one we've heard you want to sell."

We waited a bit more.

"It's being used as a safe house by the party. So the orders from Santiago are to ask you not to sell it. We would be losing one of our best assets. This may be a shock to you, as you didn't know that—"

But we did, we did know what it was being used for.

I had arranged in August of 1974 for Jean-Pierre Clerc, a journalist and friend from *Le Monde,* to visit Chile and secretly interview one of the leaders of the Resistance, Jaime Gazmuri, the secretary general of the MAPU in clandestinity. The media had been full of depressing tales of death and agony leaching out of Chile—not a bad tactic to isolate the dictatorship and get it condemned in all manner of international bodies, but missing had been news of how, underneath the conspicuous country of terror where Pinochet seemed to exercise total control, a second country of defiance was growing.

With the other heads of the proscribed parties of the Unidad Popular in exile or in prison, Gazmuri was the last leader of our political coalition to remain underground in Santiago, the last leader left in Chile who, the day before the military takeover, had met with Salvador Allende. *Le Monde* knew that it would be a scoop to speak to someone who was resisting under the shadow of the death squads. But the editors also wanted to know if we could guarantee the safety of their correspondent.

"Jean-Pierre is as safe in Santiago as he would be in Paris," I said. "We have hundreds of cadres dedicated to making sure the secretary general of our party is secure." It was a bluff, a blind leap of faith. I had no idea if this operation involved three militants or the hundreds I had invented.

It turned out that Jean-Pierre completed his mission brilliantly. The interview was published on September 11, 1974 — the first anniversary of the coup — on the front page of *Le Monde* and then syndicated worldwide. Rereading that report thirty-six years later, I'm impressed with Gazmuri's foresight. He might be under deep cover, our party leader said, but the way to overthrow Pinochet was mostly from above ground. The Resistance would start occupying — no matter the cost — the surface of the country, strangling the dictatorship with thousands of initiatives, exercising democracy in everyday life and activities, building a coalition of parties that would include the rival Christian Democrats, many of whose members repented of having helped to instigate the coup.

A few days after Jean-Pierre's return from Chile, he invited us to his home for a debriefing session.

It was good to see Santiago again, if only through Jean-Pierre's eyes: the bizarre normalcy that suffused the city, as if nothing were amiss, no torture chambers, no secret police. A woman from our party had contacted him, set up a rendezvous. Wedged into the back seat of a car, where he wore opaque glasses that blocked his vision, he had to change cars three times before arriving at his destination.

Our friend would stop once in a while to depict a member of the network and then ask us, "Is this anyone you might recognize?" We would press him for more details. My eyes would meet Angélica's, wondering silently if maybe that person could be . . .

and then we both let the name dangle, did not dare mention it, but her eyes and my eyes were saying yes, it was somebody we knew.

"As for Gazmuri," Jean-Pierre said, "I've been asked by your comrades not to describe him, he's changed his appearance significantly, but let me tell you this: Pinochet doesn't have to get him—cigarettes will do the job. He smoked nonstop during the three hours of our interview. Must have gone through several packs."

"Can you describe the house?"

"Unpretentious, with only a front yard, I think, not very large, though I could barely see through the window, but there was the most wonderful, what do you call it, jacaranda tree outside the front door. And the room where Gazmuri was sitting on a couch, it seemed a library. In fact the whole house was like a secondhand bookstore. There were books everywhere, not a wall without a bookcase. Even in the kitchen—the woman who acted as our hostess, a dark-skinned woman, slender, with long black hair, matched by eyes of the same color, well, she invited us into the kitchen to have some coffee—even there I found books." He looked at us quizzically. "Maybe you've been to that house?"

Angélica touched my foot under the table.

"I don't think so," she said.

"Tell them about the bathroom," Jean-Pierre's wife chipped in.

"Oh, the bathroom. It was painted all in orange. With a gigantic poster of Bob Dylan, the one with his hair like a rainbow in flames."

"Not anywhere I've been," I answered. "I'm sure I'd remember a bathroom like that one."

We had painted that bathroom orange ourselves one hilarious Sunday in Santiago, and the hostess in our own home was none other than Angélica's sister Ana María, who was living in our bungalow on Vaticano, and the woman who had first contacted our journalist friend seemed to be Antonieta Saa. We vaguely knew that both of them were involved, but until then we had no inkling that Ana María was leading a double existence, like in one of those movies from occupied France during the Second World War. It worried us, naturally, but at the same time it took some of the edge off our tribulations in Paris to think that our house was being used by the Resistance and made us reconsider whether its sale was absolutely necessary.

Angélica did not hesitate. That transaction had been her idea, and she was equally clear about what to do now that it was needed for the struggle, the house of our love where we had lived the three years of revolution.

"We won't sell it then," she said, ready as ever to sacrifice her own comfort for the good of others.

The party was gratified, but offered no help, no assistance to our beleaguered lives now that we were unable to dispose of the one possession that could have alleviated the financial insecurity we were suffering. And yet that Good Samaritan impulse to sacrifice our well-being in Paris ended up being a windfall, a rare instance of virtue rewarded. If we had gotten rid of the bungalow on Vaticano Street back in 1975, it would have gone for a pittance. But by the time we were forced to sell it seven years later, when we faced circumstances in Washington more dire than those in Paris, Pinochet's regime had created a new class of upstart entrepreneurs driving the real estate in Chile to levels as absurdly high as the dollar was low. So we were able to exchange our small bungalow in Santiago for a more lavish residence in Maryland, which transmogrified, in 1986, into two houses: one in Durham and, with just enough money left over, another purchased the same year in the Zapiola condominium in Santiago. Behold, the wonders of speculation and capitalism, helping us to join the middle class in spite of all my efforts to the contrary: right now, as I write this, we possess two homes instead of one, a washing machine in Santiago and a washing machine in North Carolina, a dining room table here in Durham and another one still there in Santiago, cutlery in both places, and, naturally, two mattresses.

Everything still double in my life.

But only one Angélica.

Only one Angélica.

AND NOW heaven must give way to hell, now a story of what losing that roof above your head can do to you.

Murderous thoughts.

There is a rage I need to admit to, something those fragments from my 1990 diary have thus far managed to avoid, the murderous thoughts boiling inside each time I return, threatening to erupt as soon as I step on Chilean soil.

Including the visit in 2006 when we came to film my life. No camera could capture what was seething in me as I stood in line for immigration and then customs. Peter Raymont and company could not begin to guess at the morass in my mind as I witnessed the return of other Chilean voyagers, as I listened to them chatting about shopping sprees in Miami and the delights of Disneyland. I did not reveal my scorn as I overheard a conversation between two arrogant businessmen snickering about how they fucked up their Argentine or Paraguayan or Peruvian counterparts on this last trip, they had showed *esos huevones* who was boss. I especially disliked the presumption that *nosotros somos los mejores,* we're the best, the tigers of Latin America. I am surrounded everywhere in Chile by this breed of fattened interlopers, I'm appalled by their degrading comments about the pretty young brunette who stands at the duty-free counter offering a deal on Johnnie Walker and how they'd like to screw her a few times, *tirársela bien tirada,* and teach her who's boss.

This indignation that swamps me whenever I enter my country is all the more distressing because it interferes with the joy of recognition that greets me, assures me this is where I belong, this is what I should never have left, the jokes of the baggage handlers, the courtesy of the customs officers, the bewitching smiles of the women, the smell of bread as soon as I cross into the terminal, wafting into me, anticipating the affability of a country not entirely ruined by the Pinochetista *nuevos ricos* who were able to come and go as they pleased during the years I pined for my friends. It is to thoughts of my friends, the true guardians of my dreams, that I turn as the best remedy, the antidote to the maelstrom of fury I feel at the marauders who have plundered and despoiled my utopia.

That's why, for the 2006 filming, I asked my oldest childhood buddy, Queno Ahumada, to accompany me on the trip from Argentina to Santiago. He was working at the Chilean embassy in Buenos Aires and I wanted him close by, perhaps as a talisman and shield against the turmoil that invades me when I land in Chile, good to have Queno around to remind me of happier times.

The 1973 coup did not only mean losing the country, losing the revolution, but also the loves of my life, the kaleidoscope and constellation of friends bonded even closer together by our shared experience of the Allende years.

There is nothing quite like it, the thrill of being present at the birth of a new social order when the tired conventions of the past are swept away. Nothing quite like it, to be alive when everything is called into question, the way the state operates and cities are built and children are educated and bodies make love and art is expressed, there is nothing quite like it, to feel reality crack open while you reinvent the ground under your feet, nothing, nothing like it, throwing caution to the wind, only possible because, as you risk death, a hand and a body and more than one are nearby, a brother and more than one, a sister and more than one, ready to die by your side, die instead of you, die so you can live. It is what warriors feel in the trenches as the bombs fall, that sense of immortality in the flower, eternity in the fire, a quickening of life when it is most imperiled, everybody waiting at the barricades, the real barricades in the street and the nearby barricades of the mind, there is nothing like the love you feel for that man, that woman who joined you on the road to paradise, what may turn out to be, if we are unlucky and the enemy wins, the road to perdition.

And then one day it ends. One day we were all celebrating life and each other and the next day we were all, all of us, every last one, being hunted down.

Our band of brothers and sisters was supposed to have met for Angélica's saint's day, September 12, 1973. Almost as if we wanted one last chance to say goodbye, one last final touch of the fingers, one last look into the faces we wouldn't see again for many years, to gather the tribe and dance to the Beatles, *Boy, you're gonna carry that weight,* dance one last dance before we were scattered forever, before we would have to carry that weight a long time.

Madness. To plan a party in a country on the verge of a civil war, squeeze out scarce hours to call up each friend, organize the food and revelry and music in spite of the extinction approaching, organize it probably because of what was approaching. A party was not, of course, how we spent that September 12, when the junta decreed that anybody who defied the all-day curfew would be shot. A fitting symbol of what our lives had become. Instead of a million of us trooping down the Alameda shouting our support for democracy and justice, we were strewn through the dismal city, leaving behind the time of harvesting and entering, those who remained in Chile and those who went away, the time of exile.

Exile. A telephone booth in Paris, that was exile.

We had been in Paris for almost two years by then. A story kept circulating among refugees about a miraculous telephone booth, a pay phone somewhere in the city that would suddenly break down and allow you to call anywhere in the world without dishing out a centime, a magic pay phone that kept changing places, one day in the Place de la République, the next in Montparnasse or Clichy. A fantasy, I thought in 1975, concocted by the collective need of a city brimming with lost souls. I had heard about Brazilians who would listen to a soccer game back home, relatives holding up the blaring radio to their receiver, and then the screams of *gol! gol! gol!* while a gaggle of irate Frenchmen waited to make an urgent call. Eduardo Galeano, the Uruguayan writer, once told me about a compatriot who had perfected a diabolical instrument, a coin attached to a wire that went down the slot and was brought out again and again, ensuring perpetual communication across borders and time zones. And there were hordes of Haitians who always managed to turn up as soon as the damaged pay phone materialized, and spent the next few hours greeting their impressively numbered kin, one by one by one, hogging the phone until its fairy-tale powers were disconnected.

Obviously, when I chanced upon one of those heavenly contraptions, in front of the Jardin du Luxembourg, I hadn't brought my address book with me — typical expatriate luck. I asked a fellow exile, Sergio Spoerer, who lived on the neighboring Rue Toullier, to occupy the booth while I rushed home to the other side of Paris. I'd call my parents and Angélica's mother and then, yes, each of my friends in Santiago, first of all Queno. I prayed he'd be there to answer, sing "Where Is the Life That Late I Led?" from *Kiss Me, Kate,* made absurdly apt by the new circumstances of our life. When I hurried back, quite *à bout de souffle,* breathless, Spoerer was tenth in line, behind a host of excited Africans. It would be hours before those free services became available. Just in case, I went by that afternoon. A notice had been posted: *Hors de Service.* Queno would have to wait till we met face to face.

As the years went by and that meeting did not materialize, Queno became a sort of secret connection to Chile, someone fabulous who had stored away inside him all my lost memories, prodigiously intact. Perhaps Queno more than any other person fulfilled this role because of his amazing ability to recollect the most orphaned events, anecdotes, acquaintances, school days, chess gambits, limericks and lyrics, and every stanza of every song from every musical, even the most obscure. Queno remembered the cast of every film made, the year, the director — an endless encyclopedia recalling more about you than your own mother. But for me he was the custodian not only of our personal past, *nuestro Quenito*, but of Chile itself. Not meant as a mere simile. During the dictatorship, Queno became the main archivist at the Vicaría de la Solidaridad, our central human rights organization. As such, he registered every moment of dread, every moment of hope the country endured, he became the depository of what the rulers of Chile were determined that we forget.

And then there's Cacho Rubio with his eternal mustache. Quiet, reliable, sweet Cacho, yet capable of sudden fierce violence. One evening, in a past so far away it seems as if it happened in another galaxy, we were marching in the streets of Santiago in defense of the Unidad Popular government when two thugs attacked me, *Judío de mierda, judío traidor,* we're going to kill you, fucking traitor Jew. As always when I'm the object of aggression, I was paralyzed, astonished that anyone would want to hurt poor little gentle me. As I waited to be beaten up, I saw a leg flying through the air and into the first assailant's chest and it was Cacho and then his karate chop downed the second goon, a memory that kept me warm in exile, recalling Cacho and his mellow smile when he'd come with wiry and mischievous Carlos Varas to protect Angélica and Rodrigo while I was away for the night on party business, both of them there to protect my family because a right-wing mob shouting *Long Live Donald Duck* — it's true, absolutely true! — had thrown a brick through our window, promising further aggression as retaliation against my book denouncing Disney's cultural imperialism.

After the coup, Cacho had been given the task of structuring several regional MAPU sections, then had gone abroad for training. When some of his contacts were arrested and his covert identity was compromised, it was impossible for him to go back to

Chile. A terrible moment to abort his return, as his mother had just become mortally ill: he couldn't rush home to see her or explain to the family why he wasn't at her funeral. Cacho settled first in Rome, then in Bonn, and we met as often as we could, consoled him as best we could, and then one day in 1979, we drove in a borrowed car to Stuttgart to attend his wedding to Sabine, a German girl he'd fallen in love with. Life went on in spite of Pinochet, at times because of Pinochet: there we were, with Joaquín just recently born, our family as sole representatives of a Chile where Cacho could not set foot.

The end of exile meant the possibility of living in the same city as Queno and Cacho and my other *amores del alma*. I'd comforted myself with a story — told in every émigré community — about a mythical mother who leaves a place for her banished son each evening as she sets the table. I'd whispered to myself that Mother Chile was waiting for me, for us, my friends acting as a sort of anchor, one zone of stability that wouldn't fade away in my quicksand existence. Once we did go back, at first in sporadic trips and then presumably for good in 1990, we discovered that friends and family had arranged the ticktocking of their lives in ways that did not habitually include us. They'd grown accustomed to our absence and — even if my journal entries do not spell this out — so had I. I had come to savor a sort of convenient loneliness, reserve, seclusion. So I was glad when Queno or Cacho or any of the others dropped in unannounced, but I had also grown wary of my privacy being invaded, even by the most benevolent of visitors, I had come to value one of the benefits of expatriation from a community, the freedom to orchestrate your own agenda and calendar without others encroaching on your nights or priorities.

And yet the nostalgia for reuniting the whole group remained, it was more unfinished business gnawing at me, so when the 2006 filming trip was planned I decided to gather all my friends in one place for what I imagined would be a boisterous Sunday lunch at our Zapiola house. Like so many of my madcap plots, this one hadn't panned out: it became impossible to coordinate the disparate agendas of the whole gang. *Está bien*, I thought, then I'll re-create the magic circle through my own person, by traveling among them like a comet that dashes from planet to planet, a perfect correlative of how they continue to congregate in my memory, how I am close and then I am far.

And so I went with Queno to the old building of the Vicaría just off the Plaza de Armas, where he had received the news that his friend and coworker José Manuel Parada had been abducted and murdered, his throat slit, his body dumped in a *zanja*, a ditch, the death squads showing no originality in their methods, only in the scattered selection of their victims, I mourned the sorrows of the past with Queno at the Vicaría. Next I met up with Manuel Jofré — still bearded and burly and not a gray hair on his head — on the grounds of the University of Chile where we had both taught, though now filled with students who will never again attend one of my seminars, and later Manuel and I went on to the house where he had been living on the day of the coup, the house where our group of militants received the news of Allende's death. And then Manuel accompanied me a few blocks to the Estadio Nacional.

The poet Jorge Montealegre was waiting for us at the gates. He'd been tortured inside that sporting facility when the junta had filled it with thousands of prisoners. Jorge had never been back, never wanted to step into those corridors brimming with screams from the past, and Manuel had also refused to enter its precincts, not for a soccer game, not for anything, because he had listened on the September nights after the military takeover, heard the firing squads in the curfew of the night as they killed his comrades, and he was unwilling to pretend that we could just forget what had happened there, as so many, too many, have done.

I had stayed away myself from the Estadio Nacional, where in democratic times I had merrily cheered our team with my pals. Only once had I broken that vow, the day in March of 1990 when our new President Aylwin celebrated the return of freedom to Chile. I told my reluctant friends Manuel and Jorge how seventy thousand people hushed as we heard a solitary pianist playing, down on the green field, variations on a song by Victor Jara. As the melody died, the women of the *Desaparecidos* appeared, in black skirts and white blouses, each carrying a placard with a photograph of their kidnapped man, dead or not dead, gone or yet to come back. And then one of the women — a wife, a daughter, a mother — began to dance her *cueca sola,* she was dancing alone a dance meant for a couple, dancing with a shadow, it was the absence that was being danced, the loneliness, the multitude watching the immense solitude, which was mine, which was ours. After a moment of shocked silence, people started clapping along with the music, a savage, tender beating of palms that shared in the sorrow, we were all dancing with our missing loves of history, brought back somehow from the invisibility into which Pinochet had banished them. And as if answering us from beyond time, the Symphony Orchestra of Chile then burst out with the chorale from Beethoven's Ninth Symphony, the song of the Chilean Resistance, Schiller's Ode to Joy, his prophecy of a day "when all men will once again be brothers."

That's what I told Manuel and Jorge outside the Estadio, that this heartbreak is inevitable, it is a heartbreak that we need, and I asked them to join me in yet another communal act of mourning under the majestic Andes. The task of all Chileans who refuse to forget: to repeatedly, achingly, liberate all the zones, one after the other, that Pinochet conquered, take back every last corner of this contaminated land. Would they come with me into the stadium, exorcise it one more time? I reached out my hands, my left hand to Manuel and my right hand to Jorge, and like three children playing in a garden, we walked into the Estadio Nacional, crossed one more line that they had not wanted to cross, it was cathartic to be guided by them and to be their guide.

On that same 2006 trip, another exorcism, organized by another friend from my adolescence, José Miguel Insulza. He doesn't live in Chile (he's the secretary general of the Organization of American States), but his clout, as the country's former minister of foreign affairs and its vice president, remains unabated, so he was able to open La Moneda Palace to me and the film crew. Allowed to inspect the corridors where I had spent the months before the coup, where my life had been spared the day the palace had been bombed, I was given a chance to walk into the office where I had last spoken to Allende.

And I had a rendezvous with Cacho Rubio and Carlos Varas on one of the avenues down which we had marched in years gone by, and we reenacted—to the delight of the filmmakers and the dismay of people walking their dogs, strolling with their baby carriages, and picnicking in the despondent grass nearby—our steps from the revolutionary past, with our fists in the air and our insolent chants, as if each of us were not over sixty years old, as if we were still in our twenties and our whole existence lay ahead and Allende were still alive.

And then there was Antonio Skármeta—the great writer of *Il Postino* who has shared so many of my sorrows in exile, the one soul mate who never lost touch with me, that smile of his so immense that his eyes disappear into the creases of the face, the consolation of sitting side by side at his house in Santiago, as we had done in Berlin and Amsterdam and Paris, in Madrid and Buenos Aires and London and Cuernavaca, you name it, Bonn, Rome, Washington, Pittsburgh and Naples, even Toruń, in Poland, we had covered the globe and met up each time we could and often when we couldn't, when it was madness to travel hundreds of miles just to embrace and say hello and say goodbye. Did it matter that we were now living in separate cities if nothing could undermine our affection?

And finally, the pièce de résistance, a trip to see Susana Wiener, whom I had encountered less often than I should have on my previous returns, because she had withdrawn to Algarrobo, a seaside

resort a hundred miles west of Santiago, where she shakily survived by marketing her arts and crafts. She'd saved my life after the
coup, hiding me in her apartment. That's what we talked about, as
if the cameras were not registering our faces, the fear in her voice,
back then and now, the love in her voice.

A great asset for any resistance movement: a woman of bourgeois extraction, coming late to militancy, without previous affili-
ations, not on the radar of the secret police. Nobody would have
guessed that red-haired, supposedly flighty, supposedly apolitical
Susana, with her five-year-old hyperactive daughter Solange and
her degree in art design, was in fact a courier for the MAPU, trans-
porting the most hunted leaders to safe houses. Susana soon discovered the perils of what she had volunteered for. By early 1974
she had started to transcribe in miniature handwriting the horrors
of the dictatorship. Those tiny slices of paper were photographed
and sent abroad inside who knows what, a doll, a cigarette car-
ton, buried in the sole of a shoe, and in Paris I'd have them blown
up to a readable size and transmitted to journalists, governments,
church organizations. Without knowing that it was our own Susana who copied out what was done to the genitals, the bodies cast
into the rivers, the bones broken, what might happen to her if she
made a mistake, if somebody else made a mistake.

I stayed in touch with her during my exile years through letters, and this is what I guessed each time I opened them in Paris
or Amsterdam: this woman, one of the most valiant and generous human beings, our Susana, had been crying when she wrote
those words, on the other side of the earth eight days earlier tears
had flowed as she spoke of her loneliness. But because of the way
in which letters play with time, disembody time and stretch it, I
was locked into the moment when she had been in pain, had been
desperate and needed comfort, and yet all those days later could
do nothing to heal her. So to visit her in Algarrobo in 2006 was a
way to defeat that pain, leave behind the frustrating land of letter
writing, as if she and I and Angélica and the rest of the gang could

recover that September 12, 1973, when we were supposed to have gathered, defeat time as if there had been no coup and all these decades of wandering and heartache and torments had been unreal.

But they were real. Those decades and that sorrow were unbearably real, and no matter how hard I try, the joy I feel at each meeting with my friends cannot be isolated from the memory of what we lost. I am unable to dispel entirely the anger that often, in fact, seems to fatten on those moments of redemption, mixes poisonously with that redemption. Is my rage then inescapable, as permanent a fixture of my life as love had once been, as permanent as the love I still share with my friends, inseparably held together in my heart and perhaps in theirs, love and rage, rage and love?

That trip to the coast of Chile in 2006 certainly scratched at the question. Because there was another reason for traveling there besides spending a day with Susana. In the remote splendor of my youth—I must have been all of fourteen years old—I had on several occasions rowed out with some of my schoolmates to a particularly savage island in the bay of Algarrobo. There, for hours on end, we had amused and bemused ourselves watching the antics of a group of penguins as they courted one another, and we had come back the next summer vacation to discover a population of little penguins. But it was not in order to revive one of my most cherished memories that I wanted the film crew to accompany me. I had already taken my American granddaughters on a tour of the island the previous year and found that the birds had vanished. The land at the tip of the bay, thanks to a concession from the Pinochet government, had been occupied by a private consortium, the Cofradía Náutica del Pacífico, mostly consisting of former naval officers, led by Admiral Merino, a member of the junta on whose ships many women had been raped and patriots tortured to death. The Cofradía had cut off the outlet to the sea, erecting a wall of rocks that joined the mainland to the island so that yachts could dock in an artificial cove. The penguins were less fortunate. An army of voracious rodents were now able to cross onto the defenseless peninsula and devour the eggs, wiping out future generations of flightless birds. Just as abysmal had been the effect on the bay itself. Without a natural passage to the ocean, the currents could no longer carry the waste away from Algarrobo, rendering stagnant some of the most pristine blue water in the Pacific.

That's what I wanted Peter Raymont to film, I wanted audiences to watch me trying to reach my island, watch the guards block me, the erosion of the landscape and the erosion of my dreams in one sad shot, a metaphor for Chile and the fractures of my life. I couldn't drag the cameras back to the time when I first realized how nature had been stolen from me, from all of us, by the coup: it was when, during my first return to Chile in 1983, I hiked up into the hills surrounding Santiago and was hindered by a barbed-wire barrier with a No Trespassing sign.

The military had taken over the open fields and the path into the mountains I had used as an adolescent, they were building who knows what ugly architectural monster, to be used to train more soldiers and officers where, before my exile, birds and animals and a forest had reigned. The same predatory practices were repeated all over the country, whether in the deep forests of the south of Chile or the beaches of the north, always the same guards with rifles telling me that the land that had once belonged to all was now the domain of the privileged few, those with more money and more guns. That was the point I wanted to make in the film, make it to those who lived in lands fortunate enough not to have suffered a dictatorship: when there is no free press, no control by citizens of the powerful economic interests that decide their fate, no freedom of association or freedom to criticize and correct the mistakes and corruption of those who govern us, what ends up being intoxicated is not only the social fabric but the earth itself.

Regrettably, the footage from that part of the visit to Algarrobo in 2006, my attempts to walk on my island, the guards who refused to let me in, none of that made it into the final cut of Peter Raymont's film.

But today, four years later, the urgency of this story has only increased. The gulf of Algarrobo calls out to the Gulf of Mexico with its parallel contamination by oil, the dead penguins of my island call out to their decimated brothers and sisters, the pelicans and turtles of the United States, and to all creatures — including those creatures called humans — being polluted in far-off and nearby waters.

That is what makes Chile valuable, an indispensable prism through which to scrutinize the world. Because for so many decades there were fewer limits on the arbitrary actions of a state in thrall to its wealthy patrons, and because the resulting depredations suffered by my country were extreme, Chile serves to alert us to the dangers threatening other societies, supposedly less vulnerable but just as subjugated by corporations.

The story of the penguins may also help some readers to understand where my rage wells up from, to connect my thoughts with theirs, perhaps as secretly murderous as mine.

Is that, then, the conclusion I intend to draw, that the sort of anger I feel whenever I return to my country is healthy and admirable, that we need more of it in a world where greed and egotism and ignorance are just as rampant?

Here's the rub. I intensely dislike that resentment, those thoughts of hatred are not the thoughts I want commanding my heart, that's not where I want to lodge my identity. That's not how I want to feel every time I return to Chile — or fly back to the States, for that matter — despising those I am visiting.

And yet the story of that rage, like the story of my lost and found friends, cannot be detached from the story of my exile. One of the advantages of being deprived of your country, one of its blessings, was not having to dwell under the same sun, walk the same streets tainted by the followers of General Pinochet. A sublime gift, one that I did not know existed, that I was not even thankful for, until it came roaring unexpectedly into my life one uncanny morning in The Hague.

It was sometime in early 1976, when we were living in Paris and exploring a possible transfer to Amsterdam. I welcomed any chance to journey to Holland — in this case to attend International PEN's annual conference. I'd been invited by our friend the poet Ankie Peypers, who had conspired with Frances FitzGerald, the American delegate, to introduce a motion to expel Chile from that organization. A measure not adopted since the suspension of Nazi Germany. The Dutch PEN Center had financed someone in Chile to covertly gather a list of writers, publishers, and journalists who had been executed or imprisoned, plus nineteen who were *Desaparecidos*. Ankie had also invited several Chilean writers to the conference in The Hague, including my pal Skármeta (who had yet to write *Il Postino*) and Hernán Valdés (whose *Tejas Verdes*, narrating his descent into the daily degradation of a Pinochet torture camp, had been hailed as a masterpiece all over Europe).

Our chances of ejecting Chile seemed to increase when we heard that the junta had dispatched Jorge Iván Hübner to defend its national PEN club. He'd been put in charge of the Congressional Library after the military dissolved Parliament in 1973 and had turned the building into the National Center for Prisoners (I'm not kidding), where relatives spent fruitless hours trying to obtain information about their loved ones. Oblivious to what was transpiring on the other side of the walls of what he now considered *his* library, he'd fired anyone sympathetic to Allende and settled down to censoring offensive books and magazines.

And we had another ace up our sleeve — well, it was an audiocassette, and rather than up a sleeve, it was burning a hole in my pants pocket as I approached the Hotel Des Indes in The Hague, headquarters for the conference. The next day, just after Hübner would affirm that no writers had been persecuted in his country, Ankie Peypers would play the cassette, recorded in Berlin by a poet and script writer who had been imprisoned in a detention center in the desert north of Chile—and who happened to be Hübner's own nephew, Douglas Hübner, a witness for the prosecution who would be difficult to disavow! So we were feeling cocky.

And then, on the marble stairs that lead into the foyer of the Hotel Des Indes, I met the fascist librarian. Or rather, he met me—he recognized me as I came through the revolving door and started up the steps. Before I could recoil, he had thrust an unpleasantly warm, moist fish hand into my own defenseless hand. I heard him slobber something to the effect that we were both Chileans and he hoped we could have a pleasant relationship though we might be on opposite sides of the fence, or something equally trivial and idiotic.

For a moment I was stunned, as if instead of his hand pumping mine, he was administering an electric shock, squeezing every last word out of me. And then I jumped back, withdrew my hand violently from that soft, slimy grip, and a barrage of insults spattered from my mouth. So beside myself that I couldn't remember later what I had spat out, *fascist* was the least of the epithets, according to a Dutch author who was there: fascist and murderer and Victor Jara and Allende and Neruda, and his clammy hands, how dare you touch me, *hijo de puta!* Until I had come to a stuttering halt, my hand stretched in front of me, far from the rest of my body, as if it were diseased with some toxic waste. The writers present began to applaud, a response so unnerving that Hübner beat a hasty retreat.

He left behind a trembling, distraught, and embarrassed Ariel. It was not the outburst that disturbed me. That was a natural reaction to an enemy violating the safe and peaceful sanctuary I had constructed for myself far from Chile. What shocked me was the pleasure I'd taken in the ferocity and thrill of self-righteousness when I lost control, how I wished I had a gun in my pocket instead of an audiocassette. I did not like what I was discovering about myself, how the cruelty of the coup might have burrowed into me, how easy it is to be soiled with the hatred you are trying to fight.

Not that this discomfort stopped me from surrendering to a swell of delight the next day, when the vote was taken to oust Pinochet's country from International PEN — twenty-one for expulsion, one abstention (Scotland), and one against (Hübner had, a tad farcically, cast a vote against his own ostracism). All eyes turned to the Chilean delegate. He hadn't moved from his seat in the front row, made no attempt to do so until the chairman asked Mr. Hübner to please do the assembly the favor of leaving, as he and his group were no longer welcome at this or any PEN gathering. I was seated with Skármeta and Valdés in a section at the back reserved for honored guests, and it was gratifying to see that despicable apologist for a despicable regime shuffle down the aisle, pale and flustered, not daring to look any of us in the eye, that man who had persecuted his own librarians and lied about the mistreatment of writers, including his own flesh and blood, now exposed in all his shame.

That was justice.

My murderous anger was something else. It is still inside me, I can perceive its easy contours under the waters of my mind, it is here right now as I write these words, think about all those Hübners out there and the men who served as their instruments of torment and terror. And yet, I have learned that wrath helps you to survive in the worst of times, but it cannot help us to live well, to build a lasting peace, a lasting humanity.

This lesson began to dawn on me, I think, on that first return of mine to Chile in 1983, when I was forced to share the country with the enemies who had sent me into exile. One morning I set out from the house of my brother-in-law Patricio and his wife, Marisa, desperate to release some energy and sweat, ready for a jog, my first in our country in over ten years. I had loped through those Santiago streets back in democratic Chile, when my tight shorts and long hair were greeted by catcalls and protests against the hippie gringo intruding with his outlandish foreign customs.

The dictatorship had, paradoxically, changed all that.

This time, as I trotted next to the Canal San Carlos and the gardens snaking along Tobalaba, I met dozens of fellow citizens engaged in a similar *ejercicio,* male and female, young and old, warming up, sprinting, inhaling, exhaling, all of them with the latest gear and training shoes and bottled water, all of them oblivious to the men and women laboring at the roadside, digging ditches, repairing the sidewalk, watering the public lawns. These were jobless *compañeros* who'd been herded into the government's minimum employment plan, eight hours of exertion each day for twenty-five dollars a month. They didn't look up from their work as I dashed by, just kept moving stones, collecting debris, cutting leaves from rickety trees before they fell of their own accord. They would return to the misery of their shantytowns, to the retreating eyes of their children and the desolate stray dogs, far from the shining Santiago enclaves that Pinochet's neoliberal policies had fostered.

Where I belonged. I couldn't escape that truth then and can't escape it now, writing at my desk in my house in Durham surrounded by my books and a beautiful garden and an extensive forest: what I paid for my running shoes back in 1983 was twice what one of those unemployed persons by the wayside received each month to feed and clothe an entire family. I couldn't escape then and can't escape now that in far too many disconcerting ways I am closer to the prosperous Chileans (yes, the obnoxious ones at the airport) who have profited from Pinochet's modernization than I am to the impoverished victims toiling like members of a chain gang, the workers I had once called my comrades.

Take the line repeating itself in my head as I did my stretching exercises: "American Express. You can't go home without it." The catch phrase—recited by Karl Malden in endless TV commercials in the 1980s—was, of course, "Don't *leave* home without it." But for me American Express facilitated a homecoming rather than a departure. Without that credit card, it would have been impossible to pay for that two-week first return to Chile in 1983 with the whole family — or to sport my Nikes, for that matter. But my shoes and credit card, my Americanization, my American-Express-ization, along with a host of objects acquired abroad, absolutely normal out there, had a different meaning in Chile, signaled an affluence we could not have aspired to when my young legs first raced through those streets, a time when I barely got by on a paltry income, hadn't even opened a bank account. In spite of my initial efforts to punish myself in Paris, in spite of living austerely in Holland and by the skin of our teeth in Washington, D.C., our living standard had steadily gone up, separating us from the underclass, *el pueblo unido* I had once dreamt of merging with.

It may have been that realization, the desire to run away from that contradiction, that fueled inside me a seething contempt towards those chic, up-to-date Chileans with their imported sunglasses and ownership attitudes. I didn't want to be confused with people like them! I had come home in order to oust these denizens from power. Destroy them. Yes, I wanted to destroy them, in spite of the shoes we shared, perhaps because of the shoes we shared. I wanted to hold on to the purity of my resentment, be mad as hell, give them hell, make sure I would never surrender to the same indifference shown by those joggers to the nearby pain of those unemployed men and women. A disdain increased perhaps by the fact that I couldn't say to those affluent runners what I really thought of them, not then in 1983, not on my trip in 2006, not each time I will return to Chile in the future. I'll be civil, oh I have learned the civility I did not show to Hübner, even as I refuse to exchange smiles when I intersect the shadows of his multiple associates, shiver when I sit by their side in the same café, I just can't. A refusal that I fear will not change until and unless, unless and until they have been held accountable, every last one of them purged from my life.

And yet, consider how dangerous that attitude could be. Consider this: there had been a time when the streets of Santiago were not for jogging but for dancing. With hundreds of thousands of fellow revolutionaries I had celebrated Allende's victory in 1970 and what we thought was the start of a new era in the history of humanity. Some compatriots of ours, scared by Allende's vows to use democracy to radically alter the capitalist system, were not dancing; many privileged Chileans, the future joggers of today, were scampering abroad with their lifetime savings, provoking an economic crisis and a taunt that would become the stanza of a popular song: *Que se vayan y no vuelvan nunca más,* they should leave Chile and never come back. In our minds we had already won the battle for a future without them, up to the point that we called them *momios,* mummies, calcified remnants of a past that would never return.

Those predictions about their fate were wrong. The military coup brought the *momios* and their supposedly archaic ideas screeching back to power. It was their turn to tell us that we would never return, *que se vayan y no vuelvan nunca más,* with this difference: what we had imagined as a possibility, they enforced in the real and violent territory of history. Far from being representatives of obsolete eras and retrograde doctrines, the Chilean *momios* were in fact trendsetters, the vanguard for a tidal wave of capitalism that has triumphed globally. They have become the majority, and their lives of conspicuous consumption are what everyone, even the poor, especially the poor, dreams of attaining.

Also wrong was our attempt to instill in them the fear of their annihilation, to act as if they had no right to life, to allow them to sanction us as Stalinists bent on eradicating them. These years of exile and slaughter have convinced me that such hatred perpetuates a cycle of violence, in the mind and in the streets, that needs to be broken. Not only in Chile but also in the United States, which is caught up in confrontations and divisions and exasperations that recall the Chile of yesterday.

So no, it doesn't make any sense to shout at our adversaries that they should leave and never, ever come back. Or is this what I want, a nation of enemies? Is that the world I want my grandkids to inherit? Have I learned nothing from my countless years of defeat? Do I want to endlessly repeat the conflicts of yesteryear?

I know, I know, I know.

And yet I am not sure if the next time my plane descends into the valley of Santiago, I can't be sure if the moment I step into the terminal I will not again feel a barrage of anger conquering my life, I am not sure if I will ever be able to forgive what was done to the penguins, to my country and its land and nature and people, what was done to me and my friends.

Fragment from the Diary of My Return to Chile in 1990

AUGUST 10

After two weeks back home, I'm ready to admit what attracts me most to this land, what inspires me here, how even the misery and the anguish feed me. Perhaps I'm condemned to being a literary vampire. That's the morally ambiguous fiefdom where writers abide, all of us voyeurs, trying to turn tragedies into something unforgettably beautiful, and Chile happens to be the place where more stories and more tragedies exist for me than anywhere else, where I am familiar enough with what is being suffered to understand it and distant enough so I won't be submerged by the tidal wave of my own witnessing.

Nowhere else in the world could I have had an experience like yesterday's. Our family had crossed Santiago, guided by loyal María Elena, who has been faithfully helping Angélica's mother all these years and is now assisting us a few days a week with our settling in. She took us to a remote site near her *población* that is being excavated by order of a judge hoping to find bodies of the Disappeared. Kids were flying kites as if nothing could be more natural, and athletes were griping because the bulldozers had destroyed their *cancha de fútbol* and they couldn't play their game of soccer, the perfect image of a

Chile desperate to go on with life on the surface while secrets and bodies fester under the earth.

After watching this spectacle for a while, we repaired to the ramshackle bar El Sportivo. It was Sunday, so the place was packed, and the owners had provided entertainment: a duo at the very back, one playing a guitar, the other an accordion, belting out Mexican rancheras and treacly boleros. Nearby was a short, stubbly man with a toothbrush of a mustache barely gracing his upper lip.

Once we had purchased our drinks at the counter, he hobbled towards us, holding an improvised stick in one hand and with the other putting on and taking off dark glasses, but not too quickly, and as he passed us I wondered if he was eyeless, that's how out of place those glasses were, but they were just a prop, because he kept looking back at me intensely.

Soon enough he returned, stopped at our table, and said that I didn't remember him but that I would, I'd remember him when he played the guitar. I was intrigued enough to begin to respond, but Angélica cursorily told him no, she'd never seen him before. But he wasn't interested in her, he went on and on about the day when I had listened to him and his sister, this sister wasn't here now or I'd remember her for sure, she'd married bigamously, *un matrimonio fingido,* he said, a fake marriage, and he spoke of the bells of a church that were chiming right now, right now.

I was sure it was a con of some sort, but I was fascinated nevertheless, even more so when one of the musicians handed him a guitar and he played an intricate rendition of a Paraguayan tour de force called "Pájaro Campana." He must have been a marvel once upon a time, but now his fingers were all gnarled, they had been mangled, he slurred the words to me later, in a *cogoteo,* some bad guys had beaten the shit out of him, and on that occasion or another one (it was hard to make much sense of what he was saying, not clear if he was astute or drunk or both), he had been stabbed in the leg, and now he didn't have a guitar, but *yo toco porque así soy, porque la música es lo más importante, para que nadie piense que no valgo, así como me ve,* I play because that's who I am, because nothing's more important than music, so nobody can think I'm worthless, what you see is what you get. Then he leaned towards Rodrigo and breathed at him, *putas que soi lindo, si fueras mujer me acostaría contigo,* fuck you're so beautiful that if you were a woman I'd go to bed with you, and by then, fortunately, Angélica had hurried off with Joaquín and María Elena, and it didn't matter anymore if he was trying to extract some pesos to get extra-smashed, he was a philosopher, eggs are round, he said, but human beings are rounder.

He asked for money and I gave him some, and it was the right thing to do. I had ventured into his territory with my nice clothes and my car and my family and my full pockets and he was asking for nothing more than *peaje,* a toll for passing through, and then his hand scurried inside his scraggly jacket and whipped out something small and dark, a domino, and even in that splinter of a second when I thought, It's a knife, he's going to gouge me, I also blessed myself for being so malignantly alive, blessed him for having survived all these years and for lying in wait for me in this bar. With all his dignity and all his debasement, he was bringing me a message from the seething sad Chile that I could find only here, he was telling me that the heroic and martyred singer Victor Jara is dead and what is left is this musician, with his fabled bigamous sister and his shattered life and news that he doesn't know he's delivering, only here could I allow myself to be conned so blatantly, enter into a pact where he received pesos and I received this story, only here could I belong and not belong just enough, know the code and ignore the code, only here might this happen to me.

Not a fluke, this encounter. So many others, every day.

There's a man, gaunt and leathered, with eyes so deeply set in his face you can hardly discern them, a man who "cares" for our scruffy Peugeot each time we park at the small supermarket on Avenida Larraín, meaning he motions with his hand this way or that way when we arrive or depart and gets a coin for his troubles. Some distress in my eyes, or maybe the deference with which I treat him, must have disclosed that I'm a *compañero,* that once upon a time we strode together through streets we thought would always be free, but he can't say *compañero* to me because the dictatorship has taught him not to use that word, and he won't say *señor* to me because it would indicate that we are no longer equals, not even in his and my recollection, so he has found the only word that lets him keep his memories unsullied, he calls me *amigo, gracias, amigo,* he says, I'm his friend, attempting an impossible compromise between the joyous past and the squalid present. And each time I hear that word *amigo,* it appeases a sorrow that has been mounting in me at the incessant sight of hordes of other *cuidadores de autos* teeming across Santiago, all those idle men and boys standing on "their" corner watching over someone else's property, pretending to be part of some rational hierarchy because they wear a ridiculous official-looking bailey hat with a small brim, their useless, barren, nonproductive activity valuable only inasmuch as this degraded Robin Hood scheme, taking crumbs from the rich, keeps the poor from roaming the city, thieving or begging or vandalizing cars. It's protection money you dish out each time you park, a sham, because the *cuidadores de autos* are counted as gainfully employed in government statistics, one more example of our hypocrisy. And one more reason to feel I'm to blame, my efforts to change the unjust world have been as circular and unfruitful as what these caretakers worthy of Sisyphus do in the apathetic city, perhaps this former *compañero* who calls me *amigo* will challenge me to ditch the automobile one afternoon so I can invite him to share a drink and a tidbit of his life, and maybe I'll offer him something from my own wanderings.

So that's what I'm doing here. While judges excavate for bodies, I excavate for secrets in the vast pillaged wasteland of Chile, our country an overflowing reservoir of men and women desperate with stories that have not yet been told, that must be told, the only small victories they are allowed. I have tried, ever since the coup, to keep my promise to the dead, and now perhaps I am starting to make a different sort of promise, to the living, that I will eventually, once my current period of silence is over — but how hard to stay quiet when everything around me is crying out for expression — yes, someday soon, maybe next year, I'll birth those stories into the world, extract them from the land I have defended from afar, and perhaps, just maybe, I can become the place, my literature can become, yet one more time, the place where the living and the dead meet.

•

GUILT!

Ah, what would the story of exile be, any story of survival, without guilt?

The roots of my own guilt claw deep into me, commence in the mists of infancy, though its sting was to be aggravated by the consequences of the coup, that's when I felt the true bite of its misery.

My consternation as a child at the injustice of the world cannot be attributed only to the exemplary lessons and compassion of my left-wing parents, important as their own awareness of inequality was as I grew up. It came from some well of empathy inside. I was the sort of kid who goes out of his way not to step on a solitary ant and apologizes to a tree if he happens to mark it with a stone. Whenever I suffered some injury or inequitable censure (and my ceaseless energy and rebelliousness got me into more than the usual trouble), I can remember pledging to myself that I would not forget this when I grew older, that's what I vowed, *I will never forget what it means to be defenseless and feel abandoned, I will keep inside me the memory of being treated unfairly.* It was a promise I somehow managed to transmit over and over, as if it were a baton in a relay race, to the next incarnation of that self answering to my name, so by the time I was old enough to read and study and engage in activities that might make the world a bit less cruel, the conditions were there to turn me into a fervent revolutionary.

My desire to transform society was fed, as is often the case with so many rebels, by discomfort with my own privileged existence. My family personified the typical success story of immigrants to the New World: grandparents fleeing Eastern Europe at the dawn of the twentieth century for prosperous Buenos Aires and their children doing even better. My dad had written the first books on Argentine and Latin American industry, a blueprint for how misdeveloped nations could catch up to the Western powers. In 1947 he became one of the founders of the Economic and Social Council of the United Nations. A modest Studebaker, a rented house in Queens and then an apartment overlooking Riverside Drive in Manhattan, once in a while some help from a cleaning woman, all the books and records anyone could lust for, outings to the Museum of Modern Art and the Metropolitan Opera and Broadway shows, vacations to Cape Cod and Fire Island, an eight-month trip to Europe, home leave to Argentina every other year, traveling on opulent cruise vessels.

All quite wonderful, but nothing compared to what awaited us when, fleeing the Red Scare of the fifties, Adolfo Dorfman absconded to Santiago. I was able to attend the Grange, the most exclusive school in Chile, perhaps in all of Latin America, and had friends galore and long *vacaciones* at a rented beach house and parties and rock 'n' roll. In Santiago we had the first house my parents ever owned, a five-bedroom chalet where I spent my youth—with a permanent cook and an obliging handmaid quartered in rooms over the garage, plus a seamstress once a week.

And lavish meals under the *parrón* of intertwining grapevines, and half of a small basketball court skirting a garden fragrant with orange and lemon trees.

And a back wall covered with vines.

That wall is the only thing that remains of a house that I still conjure up as my quintessential home. Destroyed, bulldozed, broken to rubble — as I discovered on my first return to Chile in 1983, when I eagerly took the bus on my second day back in Santiago and alighted at the stop, the very stop where so many years before Angélica and I had decided, almost nonchalantly, that it was time for us to get married. I got off the bus that September day of 1983 and walked along Los Leones Avenue to Traiguén Street and turned the corner and . . . instead of our chalet, what rose before me was a ten-story apartment building. Dozens of well-to-do families now lived where I had dreamt of girls with tight pants and novels without end and epic battles against injustice, where every night my mother would come to tuck me in, where every morning I would talk with my dad before the day opened its flowers. Gone, gone, it was gone, the house of my yearnings and aching lust and first scribblings and sweet convalescence from hepatitis, leveled to the ground.

I felt inanely responsible for that loss, as if I could have really remained behind to defend that house when the wrecking crew had come for its bricks and its memories. As if I couldn't accept that that's how things are: when a dictatorship takes control of a country, it also takes control of your past. By the time you go back, it has crushed the terrace where you gambled at cards into the night with your schoolmates, the living room where you danced cheek to cheek with a girl you would never sleep with, and oh those budding breasts you dared not touch, the attic where you stored your old toys in the hope of bequeathing them to children and grandchildren, all of it now flattened and trampled.

On every subsequent visit to Chile I compulsively slinked by that apartment building, watching it squat on the debris of my yesterdays. I had strolled by like a burglar casing the joint, not really ready to steal anything other than a look. I had never once risked the pain of passing the gate where a custodian would demand an identity card, I didn't want to ask permission to step on the splendid premises of my ruined past. Until the 2006 documentary posed the challenge of going inside to discover if even the whiff of anything remained. Maybe it was time to put this phantom to rest, the mother and the father who had bought that house and nurtured me in it and stood with me and Angélica and the family in our every hour of need, my parents were gone as well, ashes in two urns in Buenos Aires. It was now or never, with a film crew to protect me and force me to articulate what it meant, how that space that had been so fondly mine could turn into something foreign and cold and distant.

Except that the wall was there, the back wall of our property. For some reason, the architects had given the order to preserve it unharmed, still adjoining what was now a hideous parking lot. The same vines, the same bricks, the same colors that had greeted me when I was a lost boy of twelve yearning for New York and was still greeting me two days after the coup when, already a young man of thirty-one, I sneaked by to say goodbye to my parents on my way into clandestinity, really on my way to exile, really coming to bid farewell to the house I would never see again.

Really saying goodbye to the life of ease and pleasure that we had somehow thought would never end, could coexist with the revolution we believed in.

That such a lush existence contradicted my parents' socialist doctrines had become apparent as I grew older in that very house and conscious of the vast *llanuras de pobreza,* the poverty stretching everywhere around us, in nearby slums and haciendas. I realized that the breach between my life and the life of the proletariat I presumed to represent, and whose impoverishment and exploitation I dreamt of abolishing, was not going to be easily negotiated.

The son of one of our Communist friends—I'll call him Tony— had found the inconsistency so intolerable and hypocritical that he had moved, in the early sixties, to a shack in one of Santiago's *poblaciones*. A month later, sick with typhoid, Tony was being nursed back to health in his mother's snug apartment. My father used the occasion to drill a sobering lesson into me: "The poor don't want you to be miserable like them; they want a chance to live decently, to have some say in their life. They want you to help them eliminate the conditions which created that misery."

Judicious words, but also complacent ones, because no matter how much I worked to eradicate the agony of *los pobres de la tierra*, I knew that at the end of each day the disparity would still be there and I could, of course, withdraw to my sheltered existence.

And then Allende won the elections and for the next three years whatever dispensations I might have enjoyed became irrelevant. Our revolution was going to flush them down the toilet soon enough, not just an empty promise as we began to break down each and every barrier, as the workers of Chile on the march—my new and endless community—dispelled the inner wilderness of my guilt, abolished day by day the chasm separating me from the majority of the human race.

When the coup shattered that dream, it also brought with it poignant proof that my identification with the insurrectionary masses had become so genuine that I found myself in real danger. If I had died a martyr of the Chilean revolution, if I had not been spared death by Allende's side at La Moneda, that would have been my epitaph: here lies someone who believed so ardently in the cause of freedom and social justice that he paid the ultimate price.

But I had escaped the epitaph, and the price I paid, during those first years of exile, was—quite deliberately—to suffer as those in Chile were suffering, as the poor of Chile and the rest of the planet had suffered over the centuries. This was the secret reason I did nothing to climb out of the deprivation I found myself in after the coup.

Looking back on those years, I don't believe that in the wake of such death and collective suffering it would have been emotionally possible to avoid that plunge into the swamp of dispossession. Already fighting the drumbeat guilt of having survived, I needed at least to allay the more insidious underlying guilt of living a pleasurable life. But I wonder if my persistence in mortifying myself cannot also be attributed to something else, to an unconscious understanding that this brush with destitution was what I really craved, both as a human being and as a writer. They were the worst years of my life — and yet may also have been the best thing that ever happened to me. Being a third-class migrant, fearing the cop who stopped me in the street, not knowing if we'd have enough money to buy food tomorrow, all of that, even the trauma, even watching the woman of my life tottering on the threshold of indigence, had leveled me, taught me more than a thousand books, gave me access to those human beings among us who, on a permanent or transitory basis, are the outsiders and losers, those men and women who inhabit an agonizing global plateau of our reality. From exile I received, along with its trials, the gift of seeing the world anew, ensnared inside the skin of those who live steeped in crisis, on the edge of homelessness.

In Paris, I joined the human race.

But if exile bestowed upon me a connection with the pariahs of our time in a way that could never have happened if I hadn't been denied a country, it also in the long run confirmed my distance from those pariahs. Along with teaching me what it means to be helpless and estranged, exile also, precisely because it pushed me to the brink, verified the ambiguous sources of power available to me in my body, in my education, in my class, in my two languages, those possessions which most outcasts do not generally count among their blessings.

All this came to a head — at least that's how I recall it, this is the incident that has stuck in my craw, that I keep repulsively dragging out of the gutter of my past — on a train I took one freezing day in the winter of 1976, what was to be our last winter in Paris, a train to the working-class suburb of St. Denis, a fifteen-minute ride from the Gare du Nord, that's where and when I hit bottom.

I was in a hurry, too much of a hurry as usual, heading for an old building loaned by the French trade unions to the banned federation of Chilean workers, the Central Unica de Trabajadores. We were planning a rally for I don't know what anniversary, and I was expected to provide scraps of a libretto, perhaps convince some singers to perform, a brigade of artists to paint a mural, something like that. One more act of the many we organized abroad, staples of the diaspora, occasions to bring together the faithful, to threaten the enemy from the remoteness of our raised fists, a way to collect money — but above all to keep our community busy, feverishly useful.

That day I was late, barely managing to swing onto the train as it was about to leave the station. I had spent the morning in a café, just off the Pantheon, with Regis Debray, sometime guerrilla with Che Guevara and reincarnated as acolyte and confidant of François Mitterrand, catching up on what he might have been able to do (nothing, it turned out) for Pepe Zalaquett, who had been arrested by the secret police and who at that point in early 1976 was languishing inside Tres Alamos, a concentration camp near Santiago.

The most brilliant law student of his generation, Pepe had been the head lawyer and one of the founders of the Comité Pro Paz, created by an ecumenical group of religious leaders a few weeks after the coup to deal with the human rights abuses of the junta. A legendary figure, a giant of a man, both literally and metaphorically, he was the first person to visit the detention centers of the dictatorship, saving hundreds of lives, defending prisoners when they were dragged before the Consejos de Guerra, the notorious military tribunals. Hated by the right wing in Chile because, during the Allende years, he had been in charge of expropriating large haciendas and handing over to the *campesinos* the land they had toiled on forever, Pepe had been able to accomplish his human rights work under the junta because of the support of Raúl Silva Henríquez, the cardinal of Chile. Those years of being virtually untouchable ended when the cardinal was forced by Pinochet to dissolve the Comité on November 15, 1975. That very same night agents of the dreaded DINA, the regime's secret police, had come for my friend.

Pepe's captivity filled me with more than my usual zeal. For years we had been inseparable. We had met in 1958 at a high school chess tournament, and soon we discovered a mutual interest in social justice, painting, literature, tennis, and above all music (he could belt out the whole score of *La Traviata, Otello, Le Nozze di Figaro*). Pepe had started courting his future wife, Pía, at about the time I had begun to date Angélica, and we shared dinners and dances and *Sergeant Pepper's Lonely Hearts Club Band* and discussions on Paul Klee and so much more. His daughter Daniela and our Rodrigo were born a few weeks apart, and camping trips and vacations together and then the bonding years of the Unidad Popular all followed.

And now he was behind barbed wire and I was in Paris and there was nothing I could do to set him free. My incessant lobbying seemed to be having no effect.

So the chill inside me that morning did not come only from the biting wind and my lack of gloves. But life likes to find a way to cajole and coddle us when things seem most rotten, hand us some unexpected gift, and the train compartment I clambered into was pleasantly warm and, for once, not smelly or overcrowded. Why, there was a plush seat next to the window.

Three other passengers already ensconced in their own seats stared at me with an air of entitlement, the same pecking order and territoriality that so many migrants indulge in. I had seen fellow Chileans lord it over those who came later to their country of exile, forgetting that the whole damn place didn't belong to any of them. *Está bien*, everything would be fine on this train, as long as I didn't open my mouth, as long as I didn't let on that I was a friend of the notorious Zalaquett. I rather enjoyed my unkempt, frazzled look out of a Jean Genet play, was pleased that they couldn't bear the insolence of my crumpled clothes and unruly hair. The hell with them and their coiffed existences. I sat by the misty window, reading *Le Monde* like any literate Parisien, marshaling the heat so that enough would remain in my bones for those blocks I'd soon have to walk to the trade union federation house.

An inspector strolled in. I dispensed my ticket to him absentmindedly, absorbed in the news. Then he spoke up sharply and, not having heeded his words, I stammered back a counterquestion, promptly belying my French citizenship and *droit moral*. The other passengers nodded in self-satisfaction: Aha! So he's a foreigner. We thought so!

But it was not my nationality that he objected to. It was a matter of, well, class. I'd paid for second class and was occupying a seat in first!

I have never been good in emergencies. When something horrible happens, I tend to be paralyzed, stupefied maybe is the right word, as if I can't believe this is happening, as if it's a matter of waiting for my luck to change instantly. If I'm lucky, Angélica will be close by and rush to my defense, claws out and tongue ready to lash at anyone who dares to threaten me. But I was alone on that train, my lawyer was in a concentration camp, and my wife was babysitting a French kid.

The inspector slapped a booklet against the palm of his hand, once, twice, and began to write out a fine. I received it so listlessly that it fell to the floor. The inspector looked down at that piece of paper, then at me, then back to the floor. I crouched, grabbed the *amende*, and looked up at the other passengers, shamelessly seeking their help, begging for one of them to exclaim, as if this were a melodrama, Why, it's clear that the young man means no harm, he simply doesn't know our customs, *voilà*, let me pay that silly fine. I swear that I thought this might happen, some sort of divine intervention. But not a gleam of sympathy lit up the twin glasses of the ancient, starched couple in front of me — how can I recall these details so precisely, how they nailed me to the seat I had no right to use.

I paid the fine. I couldn't afford to be carted off to jail, have my papers closely examined. I was on a phony student visa, which an eminent professor had managed to parlay for me, attesting that I was writing a doctoral thesis at the Sorbonne. To complicate matters, I didn't reside at the address inscribed in my *carte de séjour*, a precautionary measure meant to safeguard the militants who passed through our Paris apartment as they came from Chile or returned to it in secret.

It hurt, that steep penalty, my stomach churned when Monsieur l'Inspecteur pocketed the money Angélica and I had scrimped and saved. But the human wound hurt more, the gratification those passengers wrung from this upstart being reminded that he did not belong, never had belonged, never would belong, in that compartment.

But here's the point. I did, I did belong there.

Like them, I had a safety net that I could recur to if things got really brutal. Just as Tony had been able to leave the wretched shantytown as soon as he became ill, so could I count on my parents to help out. They would have rushed to the rescue if our pride had allowed it, my need to degrade myself had allowed it.

It was Angélica who at last put a stop to this spree of masochism.

"You're only trying to punish yourself," she said one night in our apartment, a few weeks it must have been, perhaps a month or two, after that humiliation on the train to St. Denis. I'm not sure if by then our friend Pepe was in Paris, living with us, can't recall if he had already been released and then arrested a week later, and then finally deported. But I can remember Angélica's next words, spoken in a very low voice: "You want to suffer because you didn't die. You thought you should have died back there with Allende. Well, get over it. We don't need to live like this, like . . . beggars. Find a job that pays you enough to live on, and let's get the hell out of here."

Her hand swept that apartment loaned to us by Bidule. My wife was talking in hushed tones in order not to wake our hostess, whose romance with her young Cuban artist had gone sour. She had come back one night without a word of warning and reinstalled herself, half crazed, in her own house. Which she had every right to do, except that had not been part of the bargain. She was living with us day in and day out, interminably crooning to rubber ducks in her bathtub. It was all more than Angélica and I could stand. We had come to the end of this experiment in hardship.

Yet I did not say yes to Angélica until she added an ultimatum. Unlike in Buenos Aires when she had simply remained by my side, this time she imposed a deadline: "You know what, Arielito? You stay here. I'll go back to Chile with Rodrigo. Three months, that's how much time I'm willing to give you."

The next day, I sent out letters to several universities in Europe, telling them that I was available to teach Spanish-American literature or mass media or whatever, and I'd accept the first offer I received — and ten days later, one did arrive from the Universiteit van Amsterdam. Could I come and deliver a lecture there, see if we could reach some sort of arrangement?

And so, we left Paris.

It was tough for me to concede that Angélica was right: to leave the life of an outcast was to leave behind a self-image I had assiduously cultivated, my life as a minor embodiment of Che Guevara who subordinates everything personal to the cause of revolution. But it takes a special sort of stamina, a special sort of human being to continue on such a path for a lifetime. Without those men and women dedicated to the struggle against oppression — call them the small saints of change — the world would be far less endurable than it now is. There would still be slavery, women would not have the vote, workers wouldn't have the right to strike. To agree with Angélica and start thinking about myself, my career, my needs, was to begin the long, reluctant road towards discovering that I was not one of those saints, and to start accepting that the life of political heroism I had dreamt for myself during the Allende revolution was unraveling. I was scared, I suppose, of returning to the Ariel who, back in Chile before our *Presidente* had been elected, descended once in a while from the heights of his privilege. So easy to again become a tourist of suffering, troublesome to admit that my alliance with those less fortunate was utterly reversible.

Even so, the mirror in Paris had been whispering for some time that my true vocation was not in party politics, that I was destined to transform the world, if at all, by imagining alternatives, by shifting how we feel and think and write about it. As long as a blank page and a quiet typewriter had greeted me each dawn, however, I could suffocate that mirror, that whisper, that emptiness. How else to honor my promise to so many dead comrades than to keep faith with those who were effacing themselves in Chile so more would not die? What else could I offer up as a sacrifice than my own happiness?

Blessedly, by the time our dilemma became that dire, by the time Angélica demanded that we abandon Paris, the writing had started up again, providing me with a fallback position, something other than revolutionary activism to fill the void of my existence, something other than masochism as a way of responding to my guilt.

There were three of us in that rented van with our belongings, crossing France to Belgium and then into Holland and finally Amsterdam, there were three of us, Angélica and Rodrigo and Ariel, but we had left — I had left — a fourth constant companion behind.

I had left silence behind in Paris.

PART II

RETURNS

Just as the blind man walks close to his guide
in order not to stray or to collide
with something that could hurt or even kill him,

so I moved through that foul and acrid air,
hearing my guide keep telling me: "Watch out!
Be very careful not to lose me here."

—Dante, *Purgatorio*, canto XVI, 10–15 (trans. by Mark Musa)

Fragment from the Diary of My Return to Chile in 1990

AUGUST 17
Mi biblioteca!

How often, during the years of roving, had I not dreamt of the day when I would hold in my hands the first book of my lost library, place it back on a shelf, turn and reach for the next one, untouched during all these years, thumb it, read a couple of lines, glide into those pages and find verses read by my younger self once upon a time, and then look up as if roused from a delirium, the next volume calling for rediscovery — how often had this future been evoked?

Well, the future is now, the year 1990. That's how I've spent my time lately, unpacking boxes full of books, even though the rendezvous with my library has not turned out quite the way I imagined it would.

When we sold the house on Vaticano Street to remedy our distressed circumstances in Washington, Angélica's sister and her then husband, Nacho, packed the books and sent the boxes off to be kept by Santiago Larraín, a buddy from my adolescent chess-playing days, musical companion extraordinaire, tennis partner, and, when normality and chess and tennis disappeared for us, clandestine collaborator gathering information about the dictatorship. Santiago and Mafalda, his wife, offered to store my boxes of books in a shed adjoining their house in the hidden hills above the city.

A couple of months later, in the winter of 1982 — I was still forbidden to go back to Chile — Santiago phoned me in the United States with bad news: the Mapocho River had breached its banks, carrying with it houses, bridges, roads. A mudslide had providentially missed my friends' main residence, sweeping away, however, their rabbit hutches and chicken coops and dragging along, in its turbulent journey to the sea, half of my books as well.

I began to comfort him, as if it had been his library and not mine that was wrecked, not allowing myself, uncharacteristically, a morsel of self-pity.

After I hung up, I sat there for a few minutes in our new home in Bethesda, musing about this lack of sentimentality regarding the library that had formed the cornerstone of my life. Was it because those books had, in effect, vanished for me before that telephone call? Even when I refused to purchase a French-Spanish dictionary at Angélica's urging at the bookstalls on the Rue de Sébastopol, even when I needed a quote from a Shakespeare play for an article or a class, even when I tried fruitlessly to remember lines scrawled in the pages of the first edition of *Cien Años de Soledad,* even as I lay in the bleak nights of banishment and reviewed the exact ethereal order in which certain favorite tomes were lodged on a shelf, it took Santiago Larraín's call to make me realize that, in spite of all that longing and love, I had not really believed in the existence of that library.

So I was grateful to that river for savaging my most prized possessions. It was as if inside that much larger, more destructive flood of history that had exposed my life and left me naked, this Chilean rising of the waters had rescued the library, had oddly made it tangible for me once more. Instead of mourning the half that had been lost forever, I rejoiced at the resurrection of what I had given up for dead.

A time machine, that's what reading from my library has turned into. Because every volume I dig out of its box here in

Chile, saved from the soldiers and the deluge, offers me an expedition to the past, almost a geological inquiry into the layers of the life I used to live, a way of communing with the eyes and mind of the boy, the adolescent, and then the young man who slipped into the covers of this novel or that treatise on philosophy, meeting old friends again. Emma Bovary and Alyosha Karamazov and Aeneas and Joseph K and Gilgamesh and Electra and that old fool Polonius accompany me once more, though not quite as when I last left them; I have learned something since then about dead bodies and betrayals and ethical distress. To read Dante today, after having tasted the bread of exile, is not like yesterday, when I believed I'd live and die in the country of my choice.

Here was my Cortázar, one of my favorite stories, "La Autopista del Sur," which I read before I ever met him, before we ever sat down to listen to Bessie Smith in his apartment near Les Halles, before the day he confided in me that he was leaving Ugné and had fallen in love with Carol, before we stepped together into the sea of Zihuatanejo. I skimmed through that story again only this morning, remarking on my scraggly annotations in the margins, the same handwriting then, at the age of twenty-four, as now (some things, amazingly, stay the same), and the shock of recognition: Cortázar subjects a group of voyagers, returning by car to Paris one evening, to a colossal traffic jam, one that lasts for over a year, asks what would happen if they had to live in their vehicles and endure winter and hunger, reduced, as were the earliest humans, to nothing other than their bodies and their solidarity. What would they discover about themselves once some mysterious force threw an apocalyptic monkey wrench into the cogs of civilization, making us wonder where we are going and why and with whom? Without Cortázar's prophecy, itself springing from the nostalgia for the primitive and ghostly that informs the vision of so many anti-bourgeois artists, from the Romantics onwards, I could never have cooked up the bizarre theory of potholes that I espoused to my parents in a letter just a few weeks ago, potholes in the streets of Santiago as messengers from a mythical magical Chile lurking underneath or behind or beyond the everyday, haunting the ordinary, challenging the conventional. And yet, nearby, in my lost submerged library, my annotated Marx and Descartes, Asimov and Sarmiento, await me inside books that proclaim the need to tame Nature, I was also formed in this other tradition that stresses the importance of science and progress.

Here then are the metaphors and paradigms and characters from which I can never be expelled, a vast imaginary realm that will always be mine. It was here that I learned a long time ago, in the boundless pages of this library, to become cosmopolitan — Diogenes must be here, one of these days I'll look for the passage where he invented that word derived from *cosmos* and *polités*, a citizen of the universe, someone who does not have to fear exile because, gloriously homeless, he belongs only to his thoughts and to what is eminently human. Is that why I've come back to Chile? To dust off the pre-Socratics, to dip again into the classics, to recall that Socrates, according to an essay by Montaigne, said he was not from Athens but from the world, to acknowledge that when I first read those words I didn't really understand them as I do now, after decades of wandering. So much makes more sense now that I can visit the women protesting that their *Desaparecidos* are still without a grave and on the same day read of Antigone's determination to bury her brother.

But the collected plays of Sophocles are not just spiritual passports to the universe. They are also inside an object to be disinterred, hefty and material and, above all, dirty. The words inside may shine, but the pages that carry them are caked with the dust of their own transitory funeral. I can't read Euripides or Plato until I've wiped down the books where they reside, where they have established their residence, unless I've cleaned up the residue left behind by time and the river . . . Yes, Thomas Wolfe is also here, I read him before I had an inkling that I'd establish myself for so long in the North Carolina where he was born, I won't be able to look homeward with his angel until . . . my muscles and fingers have done some exercise.

Though not alone.

About four hours after I had started on my labors I came down with an asthma attack. The previous evening I had already been left dizzy by the noxious smog of Santiago, the third time in as many weeks. This is also *mi querido* Chile, this mad miasma welcome, the crushing headaches, the substandard gasoline, the buses belching fumes, the misguided industrialization of ages exacerbated by the anything-goes dictatorship and the public indifference to the common good, and, of course, the mountains that I love so much, the mountains hemming in that foul air, exacting revenge on the descendants of the conquistadors who came to despoil and celebrate this valley. The heaving beast in my lungs had subsided by the time I awoke the next morning, but the spores in the books ferreted their way into me soon enough, so that by noon an allergic reaction endangered the Great Rescue. At this rate it was going to take me more weeks to clean up my library than it had taken Tolstoy to write the Battle of Borodino, maybe longer than Napoleon's retreat from Moscow, which I was itching to read again in the Constance Garnett translation. I would glance at a description of Pierre watching the slaughter and then sneeze, read another line about Pierre's yearnings for Natasha and then dab my eyes, and so it would have continued if salvation had not come at the hands of a spindly twelve-year-old lad with a shudder of unruly black hair and clean olive skin.

Miguel was at our house with his father, a stonemason summoned by Angélica to cover the exposed bricks in Joaquín's bedroom so tiny spiders and bugs won't creep out and attack our son. Miguel should be in school, but I suspect he's been pulled out of the system, that his dad wants him to start working — not that he's much of an apprentice, as the kid kept floating away from the masonry job to contemplate me pounding the soil away from a binding and then absorbed in one of William Blake's poems and then *un estornudo* and a similar operation with Rilke or Lagerkvist, and I can't remember quite how or when it happened. It would have been narratively appropriate if it had been when I was trying to salvage *Oliver Twist*. All I know is that Miguel suddenly appeared with a rag, given to him by María Elena. Concerned by my health, she had instructed Miguel to sit down on one of the boxes on the red-tiled terrace of our patio and make himself useful. And so he began taking out books, scrubbing each one at a clip ten times faster than anything I could accomplish, weighed down as I was with asthma and curiosity and my hankering for a repetition of the pleasures of *le temps retrouvé*.

He's been my assistant for the past few days, a bright kid, quickly able to classify like-minded books in piles that I can then accommodate in relative order on the wooden boards of the rapidly filling study. It is not a strict division of labor in which the boy sweats and I read, because I'm also getting my fingernails filthy, but there can be no doubt that my aide-de-camp bears the main burden of this toil. He is happy as he plugs away, especially when the talk turns to the books. Besides making more money in these scant hours than he'll earn for the rest of the month (I doubt that his father pays him at all), Miguel is receiving the rudiments of a literary education, denied to him by the accident of his birth. He's avid for the stories embedded under the covers which I transmit to him from time to time as he wipes away the grime with fruition. Miguel nods, frowning with concentration when I recite some lines from Nicolás Guillén or tell him a fable by Jorge

Luis Borges — Borges whose work I love in spite of his having been decorated by Pinochet, Borges who wrote about a library as infinite as the universe but never once conjured up a child scrubbing his *Ficciones* on a sunny day in winter, never once stopped to think that the intellectual delectations of eternity and avatars could be denied to that boy because of what that very general has inflicted on our country, never understood that if this were still Allende's Chile, Miguel would inhabit a nation where his future, as a reader and as a worker, would be diametrically different.

I did not expect to be using the services of anyone like Miguel. In fact, at dinner at Antonio Skármeta's five or six days ago, my dear friend Antonio, upon hearing of my latest tribulations (I had spent a whole morning flitting to offices scattered across the city in order to retrieve a package from Nan Graham, my editor at Viking in New York), suggested that if he and I and maybe another writer pooled our resources, we could share a "junior" who could run our errands and would be better off with us than by begging on the streets or selling asparagus one day and trinkets made in Hong Kong the next. I launched into an Arielesque tirade about our fraudulent service society, its roots in the semifeudal past, why we assumed that so many of the poor were automatically supposed to do our bidding — and added that I wouldn't pander to such exploitation, my years abroad had taught me to unshackle myself from this sort of bond. A prodigious speech, but look at me now, less than a week later, with my very own private helper.

Though not for long. The library cleaning is coming to its end: tomorrow I'll send Miguel off with an extra-large tip and some books. And then indulge my fancy with other Miguels remaining behind with me: Miguel de Cervantes and Miguel Angel Asturias and *Miguel Street* by Naipaul and, of course, Michel de Montaigne.

What did he say, my Michel, about poverty and the mind?

I shuffle through the cold to the almost full library and find the *Essais,* perched close to Éluard and Rabelais. I open the volume, find the phrase underlined by a furious pencil that was once mine: "Poverty of goods is easily cured; poverty of the mind is irreparable," what I already selected as notable from Montaigne years ago.

When I first read that line I actually believed it was possible to cure the pandemic of poverty, yet here I am in a Chile brimming with kids like Miguel and adults like that man who played the guitar, that man who calls me *amigo* in the parking lot, here I am, slumming for stories, hoping to add more works of my own to this library, hoping that I can carry on the only struggle left to me in this land, wagering that the poverty of our nation's mind is not as irreparable as my dead Michel de Montaigne, now rescued from a sullied river, once wrote.

Because something is beginning to form in my own mind, something is starting to call from inside me and from out there in the country, something is asking me for the words it needs to be born.

Something to assuage the sorrow for me, for Miguel, for us all.

•

WE CAME to Our Lady of Amsterdam, that city of water and bridges and solidarity, in the late summer of 1976.

Unconcerned by reports that it was close to impossible in overcrowded Holland to find a residence that was both decent and relatively inexpensive. Hey, it can't be worse than France and all those incessantly shifting abodes of the previous years. And no need to hurry anyway. Max Arian and his wife, Maartje, had arranged for us to occupy for a few weeks an apartment of friends on vacation, up four steep, narrow flights on which Max and his sons, Jasja and Jeroen, helped by a fervent Rodrigo, scurried up and down with boxes and suitcases.

I had met Max on my first visit to Amsterdam, less than a year after the coup and almost three years before we moved there permanently. Several Dutch acquaintances contacted in Buenos Aires had mentioned him as the one indispensable person to know in the Netherlands, practically Mr. Solidarity-with-Chile in that country. Max was a survivor of the Holocaust, one of the Jewish children saved during the Nazi occupation when his mother went into hiding and his father disappeared into the *Nacht und Nebel* of Auschwitz. Not that he mentioned it during our first afternoon together, in April of 1974, when he took me down to the south of Holland in his mother's rickety car to see a play about resistance to Pinochet. I didn't give him a chance, chattering so animatedly about Chilean culture under Allende that he lost his way. As we tried to get our bearings, barreling along the highway in search of the missing troupe, we were pulled over by a police car. My friend felt me go tense with a nameless apprehension as the uniformed man approached us, and he and Max began to talk in a language I could not comprehend.

"Ha!" Max said when the policeman withdrew. "He says he thinks you are talking too much, gesticulating too much, too much excitement, you know, he thinks maybe I will crash if I keep listening. He let me off with a warning, as long as you don't tell me more about Allende and culture until we stop. Ha! This is very funny!"

We tried to remain silent — well, reasonably silent — for the rest of the trip back to Amsterdam. It was getting late, and Max had to get home soon because his wife was expecting a baby. Max was almost apologetic, he had felt bad, he said, because it had been the day after the coup against Allende that Maartje had found out she was pregnant, and that child had been growing towards life inside her mother while every day brought news of death from the Santiago that Max had fallen in love with, that beautiful girl, Adindaliesje, participating as little more than a seed in all those protest marches.

The Arians were our first hosts in Amsterdam and had our refrigerator loaded with food, and on the table sat a bottle of Chilean wine — but because there was a boycott of Pinochet's products, Max had searched out a pre-coup vintage. Next to the bottle was a Dutch-Spanish dictionary and a few lines of doggerel saluting our arrival. We felt so much at ease that the next day I came down with a rampaging fever, a throat so sore that I could have taken another trip with Max and no policeman would have stopped us to demand that I shut up — I could scarcely groan out a word.

Maartje, tall, cheerful, rosy-cheeked, pooh-poohed Angélica's suggestion that we call a doctor. Not in Holland, where people took maladies and adversity in their stride.

"How often has Ariel been ill in France?"

"Never."

"Ach, that is it, then."

She was right. I couldn't afford to get sick in France, perhaps something in me had fought against visiting a Parisian physician, who would treat me out of pity, some antibody in my system kept bacteria at bay in order to avoid further mortification. I thought back to Buenos Aires. Every time I would visit my native city from the States as a child, I had been beset by asthma, an outbreak of breathlessness and eczema that was as much due to the humidity and pollen as to my anxiety at returning to a country where I did not want to live. So I had expected that my first days of exile in Buenos Aires, in December of 1973, would be infernal. I withstood the blistering heat and dampness of that summer, however, without the slightest sign of habitual discomfort. As soon as I landed in Lima, en route to Cuba — finally safe from the Argentine death squads — my lungs had all but collapsed and I had been dashed to a hospital. Very simple, the doctor elucidated — the same opinion echoed a week later in Havana, where again I was barely able to breathe — you've stopped producing adrenaline, no longer in fear for your life.

In that way, Holland offered me the best welcome I could have imagined: a place where you can feel ill. A place where you can heal.

Though first we had to find a place, a real place, to merely live.

Two weeks later, still unable to rent anything satisfactory, we accepted a proposal from our poet friend Ankie Peypers to rent her studio flat in Amsterdam for a trifling sum. She'd remain, she said, at her main residence, the home she shared in Arnhem with her husband, Marius. We were sure it would be for a short while, but it wasn't until mid-November, three months later, that we were offered an apartment in Buitenveldert, a neighborhood on the southern edge of Amsterdam. The renters, Wim Gobets and his wife, were off to England for a year or two and would sublet it to us at the municipal-controlled rate. It was perfect: a living and dining area with ample windows and a sunny balcony, two small bedrooms, a bathroom, a narrow kitchen. Perfect, except for a slight drawback: the Gobetses didn't want to subject their cats, Bilbo and Hopje, to a six-month quarantine in Great Britain, so if we wanted the apartment, we'd inherit the two Siamese cats as well.

If they had asked us to take care of a pack of hyenas and a herd of giraffes and five elephants to boot we would have said yes. Of course the cats would go crazy if shut up for half a year; we had friends who were still rotting in concentration camps, so why not help those feline friends cheat their jailers, welcome them into our home as they were welcoming us into theirs?

Our keenness did not last.

I'd never had pets because my mother was allergic to animals, but Angélica and I had once adopted a stray cat in Santiago. When it had died of an infection contracted while giving birth to a litter of tiny mewling kittens, we'd used an eyedropper to feed milk to the babies. So nobody could say we weren't favorably disposed towards the feline species.

But oh how those two Dutch cats rankled. I resented those pets in direct proportion to the fading of our chances of going back home, as evidence poured in that far too many people in Chile were buying into Pinochet's counterrevolution, as international loans to his neofascist government increased. It was illogical and even callous, but I was offended that Bilbo and Hopje were permanent residents of the Netherlands whereas we were foreigners, intruders who needed to apply each year for a renewal of *our* residency, unblessed with any permanence whatsoever.

They were real masters of our temporary home, those cats, whose stinking litter box it was my job to clean, whose hair I had to brush from my clothes, whose claws I had to prevent from tearing our secondhand furniture, whose stench I had to inhale whenever I ventured to the bathroom. Lazy, aristocratic, privileged, their passivity mirrored and mocked our own lives as we waited for news of a liberation that never came. At those times when I cut raw beef heart for their meals — a promise to the Gobetses we solemnly kept — slicing tiny chunks from the fresh meat we bought so their highnesses could digest and shit and piss some more, reek up *their* apartment, they represented everything that was eluding my *pueblo* on the other side of the world. I fought the cliché, I detested clichés even more than I detested those cats, but there it was: pets in these overdeveloped countries being treated in ways that would drive people in our misshapen, misdeveloped lands pale with envy. Or maybe what I begrudged those cats was the debt I owed them: if not for their pampered existence, we would have been forced to become illegal squatters in someone else's building.

I am not joking.

As I cared for those cats and cursed them for scratching a sofa offered us by friends and blessed them for keeping us from the rain, I came to realize that we were not the only vagrants in these lowlands. The housing crisis was so acute that many young people, although flourishing by Latin American standards, had no other recourse than to take over buildings left vacant by owners unwilling to rent out flats at the subsidized rate. We were used to the *tomas* in Chile and Latin America, droves of rural migrants occupying wastelands on the fringes of cities, nailing up a couple of wooden planks and a tin roof and unfurling the national flag. In my more radical student days I had helped them exhibit that flag — the first thing they did, because this ostensibly gave them protection against police brutality. Whereas in Amsterdam, the *krakers* (a name derived from the creaking sound of a door being forced open with a crowbar) began by calling the electric company, since electricity could not be denied to anyone; then they installed a cot in the vacant place, because according to Dutch law, if someone spends twenty-four hours under one roof, sleeping in a bed, no matter how makeshift (as long as it is covered with a sheet), the "tenants" can be dislodged only by judicial order, subject to all manner of delaying tactics. Eric Gerzon — one of my students, who was to become like a brother to us — had lived for several years next to a group of *krakers,* who, for instance, altered or misspelled their names on their mailbox to avoid being served with a warrant. No flags. Just walkie-talkies that could rally a thousand protesters at the slightest hint of an eviction. In sleek magazines, the trespassers explained how to pick a lock, the ten ways of extending a telephone line, the best days for an occupation (the queen's birthday or the anniversary commemorating the victims of Nazism); they gave phone numbers of lawyers, nurses, and food co-ops.

I watched this spectacle with astonishment, still tied to visions of the remote misery of our *poblaciones,* the memories of the mud in which children and dogs and chickens were mired, a shack rank with smoke, a funeral under the rain. Every day we learned more about distance, how close the Dutch were to our needs, and how far. I was perplexed by this combination of something entirely illegal with a set of rigid and consensual regulations, amazed that people who could eat and drink, go to concerts, read books, and be fully employed should also be as homeless as we were — perhaps more so.

More homeless than we, those Dutch citizens? I could venture that bemused observation because for the first time since the coup, I felt that I had a home, a place to call my own, for the first time dared to believe I could be happy again. I surprised myself one morning singing as I biked to the university, up the untrammeled road along the Amstel, far from my Chile and alone with my brooding thoughts, and yet I sang. I found myself crooning Mozart into the cold air, never mind the wind capriciously against me, almost blowing me to a standstill while I valiantly belted out "Non più andrai," singing "Là ci darem la mano," and Papageno's tootling love fest and "Mi sento dal contento, pieno di gioia il cor." My heart was full of joy. Despite a dark side to my psyche, a certainty since childhood that everything is unreal and doomed to extinction, despite the sick conviction that I will be made to pay for each banquet and interlude of love, the core of my personality has always been jubilant. Bubbling over with a zeal for life and pleasure, an almost infantile and infectious *optimismo,* I rise each morning to the adventure and marvel of the day, an inveterate insomniac because I've never wanted to miss what the next instant might bring. And Holland was allowing me to entertain the possibility that I might filch moments of delight even from a world made by Pinochet, that not only a vale of tears lay before us.

Survivors have trouble coming to terms with their own vitality, each breath swindled from the dead. Life goes on mercilessly, often exultingly. A widow who has her first orgasm after her husband's execution, an orphan who startles the air with his own mirth, a banished writer biking through the Vondelpark with his off-key baritone version of the Ode to Joy—we tell ourselves that this is what our loved ones would want, to live for one, for two, for three, not shuffle off into sorrow, enslaved by the past. Pinochet wants us to be sad, so let's dance, my dear, just to spite him.

Yes, it was time to admit that I was glad to be alive, a resident on this earth of ours. Time to bring a new child into the world, to turn the tide of our lives and the tide of history in a new direction.

The second child had been postponed for far too long.

Not now, we thought, when Rodrigo had just been born and we were off for a year and a half to Berkeley, spending 1968 and 1969 among the hippies and the Haight-Ashbury crowd and jazzing it up under the stars and making love and not war, not the time to dedicate nine months and the years after to reproducing the species. And then the Allende revolution—no time to bring a baby into the world when you're giving birth to a country, a new social order, no time, no time—and then the coup and the years of wandering, not in Paris, inconceivable to even consider adding one more mouth to feed, one more body that could be sick. Only when we arrived in Holland, only after the first two years of stability in our new refuge, surrounded by so many friends and—no minor matter—with health care and insurance and a job and a bank account, now's the time, now's the time.

Not that conception and pregnancy automatically followed that momentous decision. For a whole assiduous year—with ten-year-old Rodrigo, desperate for a brother, jumping up and down on our bed, demanding that we get to it, what were we waiting for, *vamos, vamos,* crazy, lovely Rodrigo—we tried to engender that baby. Though the doctor told us not to worry, this was natural after years of using contraceptive devices, I couldn't help wondering whether something was wrong, whether this was a perfidious Pinochet black-magic Herod-like intervention to ensure that no child came into this world who could tip the balance against despots like him.

In fact, we were helped to conceive the child by the Sandinistas. Here's how: in May of 1978, I had arranged for two Nicaraguan writers to travel to Holland, Ernesto Cardenal, the extraordinary priest-poet, and Sergio Ramírez, who even then, before his subsequent masterpieces, was considered one of the great novelists of Central America (and would become vice president of his country in 1985, in the Sandinista government). The pretext for the journey was a Latin American literary conference, organized and paid for by the Danes and held in a seaside town north of Copenhagen. The real reason, however, was to put these two envoys of the Sandinistas in contact with the European political world, starting with the Dutch Social Democrats, who were eager to open other doors.

Sergio Ramírez, in whose house I had lodged in Berlin for a few days when I passed through to drum up support for the Chilean cultural resistance, did not come to Amsterdam empty-handed. His spouse, Tulita, who had heard about our unsuccessful efforts to conceive another child, sent Angélica a fertility god, a tiny gold frog bought in a market in San José, Costa Rica. My wife gratefully hung it from her neck, hiding from Sergio her frustration that my need to join him on this trip to Scandinavia was taking me away from my reproductive duties during the period when she would be most likely to ovulate.

Another month lost!

The night before I left on the Nicaraguan solidarity tour, Angélica and I made love for the first time in months, not in order to propagate the species but merely to console our loneliness. The nearby fertility frog must have been smiling away — that's how you do this, folks, the child will come as soon as you stop trying so hard, just enjoy the miracle of two bodies still alive in times like these, that's all, *nada más que eso* — because lo and behold, by July 1978 (exactly one year before the Sandinista triumph!), the small bump in Angélica's tummy confirmed what the medical tests had already told us: she was expecting a baby. And I was expecting my own baby — well, let's say I was pregnant with literature, because a novel was growing by then along with the child. Everything seemed to be falling into place, literature and fecundity and history itself.

Just like Joaquín, that novel had been a long time coming and, also like our second son, had sprung from exile, from a wild idea that had visited me a few months after our arrival in Holland, once the realization that Pinochet was defying our prophecies of his imminent demise sparked questions about the cost his reign of terror might be exacting on Chileans, especially the young. If I could only find a way to reach those forbidden compatriots . . .

Around this time, I interviewed at great length — and Angélica, with her usual generosity, transcribed the interview so it could be published — a friend of ours, Oscar Castro. He'd spent the previous years in a series of Chilean detention centers, punished for having put on a play in Santiago in which the captain goes down with his ship while auguring better days ahead — an allusion to Salvador Allende that the secret police found all too transparent. Other members of his troupe were also jailed, and his mother and his brother-in-law were "disappeared." Despite the depravity visited upon him, Oscar had nevertheless continued to be creative. In the first concentration camp, Melinka, where he had been sequestered, the actor-playwright had staged subversive comedies he'd written himself but attributed to that "famous Austrian author Emil Kan" — which the military commandant approved for performance so as not to show how ignorant he was of theatrical history.

And if Oscar could do that behind barbed wire, under the restless eyes of guards who could beat him up, kill him and yet more members of his family, what was to stop me from doing something similar from the freedom of exile?

Maybe I could write something in code, a story about a Greek village during the Second World War where all the men have been disappeared and only the women are left to wait, echoing a poem I had recently written, in which a body floats down a river, unrecognizable, battered by the rocks or by some malicious human hand, and is claimed by an old woman, crazed by the loss of the four men in her family. In the novel, I thought, when the military burn the corpse and ban the funeral, another body would materialize, surging from the unconscious rivers of the country. I'd arrange for my novel, that cross between *Antigone* and *The Trojan Women*, to appear under the name of a Danish author who had presumably died under the Nazis and whose manuscript had only recently been discovered by . . .

I contacted Heinrich Böll, and he agreed to "find" that lost text. "If I helped Solzhenitsyn smuggle his manuscripts out of his country, why not help you get yours *into* your Chile," he said with a mischievous twinkle in his eyes as we sipped tea and munched strudel in his house near Cologne. And then my mentor, Julio Cortázar, whose extraordinary novels and short stories were my favorite fiction, told me he was willing to offer his name as "translator" into Spanish from the nonexistent French. "It will be the easiest translation of my life," he said when I visited him in Paris, and he grinned and put on a Louis Armstrong record from the forties to celebrate our hoax. "I just take your original Spanish and claim all the glory!" It was an elaborate literary fraud that received encouragement from a prominent book editor in Chile, who told me he would consider publishing my novel in Santiago, under that Danish pseudonym, of course. Depending on its content, of course.

My parents would fly in from Argentina each summer and take us on a vacation we could otherwise not have afforded, and that year I had persuaded them to visit the land where I wanted to situate my swindle of a novel. And so the summer of 1978 saw me on the isle of Crete, typing the first words of the new novel on the same machine that, once upon a toilet, had refused me all access. Those first words: *Otra vez la vieja de mierda? Otra vez?* That old bitch again? The same old bitch? Words spat out by the exasperated Captain of that small Greek village. Before that week in Crete, Angélica and I rented a car and explored the Peloponnese for five days while my dad and mom took care of Rodrigo at a pebbly beach near Athens. "Look," I'd say to my wife, "look, there's the river where the bodies will appear. Look at those goats, look at that old woman in black and her unblinking eyes."

And Angélica would smile and eat an infinite amount of tomatoes and goat cheese and olives and bread baked just a few hours before, filled with a pregnancy more glorious than mine. We slept under constellations named for Greek gods and were befriended by riotous fishermen who would have made Zorba seem like a piker. We danced with them in taverns close to where Agamemnon had come back to death from Troy, and we cast questions into the ravine at Delphi and saw Oedipus blind himself at Epidaurus and harkened to the lapping sea that had brought Odysseus home. The novel kept ripening in my heart until we reached that cabin in Crete and then the early mornings of my typing, far from that miserable hotel in Paris and the night full of doubts and questions. So liberating to write a story about my country which was also the story of so many other unfortunate and neglected lands and widows and mothers left behind with rivers and captains.

I proclaimed a magical connection between the child gestating inside Angélica and the words being roused out of their slumber by the Muses, invented by these very Greeks in order to personify the unfathomable act of artistic possession. Every day that our little one grew inside Angélica was proof that life was stronger than death and that the people would rebel against the dictatorship as I had rebelled and defeated silence, every word that I wrote confirming that the energy and faith of the Allende years had not been crushed, that the future stirring in Angélica's womb anticipated victory.

Our dead would emerge, as we had, from the shadows!

And then, in November of that year of 1978, as I was putting the final touches to the novel I was to call *Widows,* the real *Desaparecidos* made an appearance. The bodies of fifteen peasants—taken from their homes in October 1973 and never seen alive again—were discovered in an abandoned mine shaft near the rural village of Lonquén.

After Lonquén, the *Desaparecidos* began to materialize in other places—washed up from the sea, rising from inside mass graves under plowed fields and hills where the army trained recruits, disinterred from the banks of rivers, corpses preserved by the sands of the Atacama Desert, a slow series of unearthings—but no bodies challenged the Pinochet regime more than those in that mine shaft. Up until then the conflict had been between the secluded, private memory of the relatives and the all too public guns and rebuttals of the military and its contemptuous newspapers. Now it was the irrefutable violence done to the skin and loins and bones of the victims themselves that erupted into the consciousness of a nation that had for the most part preferred, out of fear or convenience or connivance, to accept the false version of those disappearances. Before the dead of Lonquén could be properly buried, they were kidnapped—once more!—by the government and thrown into an unmarked grave. Again without a tomb to visit, the women began a weekly Sunday pilgrimage to the mine, the place where the crime committed against those lives was written on the very rocks and landscape.

I watched all this from Amsterdam with a mixture of dismay and euphoria. The dismay needs no explanation; the euphoria is more mysterious and tricky. The bodies had come back, I believed, because the surviving women had willed them from the land of death. But I also felt the elation of my own marginal, remote role as a voice for those missing men. I had started to write *Widows* months before one body had surfaced. Instead of depicting those violently ended lives as an absence, I had anticipated what would happen if those corpses did return, how the remains, by their mere refusal to remain hidden, would demand justice.

As if it had been my words that had conjured up the dead.

Forgivable, I suppose, the hint of giddy arrogance I can now retrospectively discover in the thaumaturgical role I ascribed to my writing, forgivable because it must have been a way of turning myself into a pale protagonist in events transpiring in Chile without me. Maybe I needed the illusion that my words were shaping the reality and the struggle itself, that I was a prophet even if I had no one to listen to me or read those words back in my own land.

What, then, could possibly go wrong?

A letter from the Amsterdam housing authority, that's what. Our friend Eric painstakingly translated the contents. Meneer and Mevrouw Gobets, the leasers of our apartment in the Buitenveldert, had announced that they weren't returning to Holland. The law allowed them to sublet the property for one year, and they had been given, exceptionally, an extra twelve months, so that by this coming December of 1978 the premises had to be vacated for a new tenant with the right to subsidized housing. Could Meneer Dorfman make arrangements for an orderly signing over of the apartment?

I looked up from that notice of eviction. There was Angélica with her already protruding belly, there was Rodrigo racing around and whooping it up as if the sky was not about to fall, there were my parents who were staying with us in Amsterdam for a few weeks after our Grecian holiday and would soon be back in the relative safety of their apartment in Argentina. And, strolling aristocratically into the room, here came the cats, Bilbo and Hopje, the veritable landlords of this apartment, soon to be dispossessed, out on the streets along with their guardians. And there, luckily, was Eric, as calm and collected as anyone who knows the language and the ropes, Eric, who told me not to worry, something would be negotiated. And, in effect, with Eric's help and the support of Dick Bloemraad, a student who had been translating some of my stories into Dutch, the municipality gave us a surcease of six months, until July 1979. But hey, we needed an extra year.

So I put on my best clothes and headed for an interview with a woman whose name I have thankfully forgotten, but who received me with deference, seemingly impressed with my university professorship and credentials, the pile of publications I brought with me in a briefcase and built up into two small towers on her desk, my hyped-up CV, my promise that if she would grant us a stay of execution, our family would be gone by July 1980. When she had all but assured me that there should be no problem in according such a distinguished scholar a year's grace, I made my final gambit. Out of my briefcase came *Avenue,* a Dutch cross between *Vogue* and *Marie Claire.* I pointed to my name in the table of contents. One of my short stories had appeared in this issue, and I wanted to leave it with her and write out a simple dedication, if she didn't mind. She purred and cooed as I scribbled, in English of course, something like: To our Dutch savior, who has offered us refuge, here's a story for your enjoyment.

We parted the best of friends.

Two weeks later, a letter arrived from the selfsame administrator, reiterating that the municipality expected us to leave our abode by July 1979 — or face eviction.

"What happened?" an astonished Angélica asked me. "Wasn't she your best buddy? Didn't you tell me you had charmed the pants off her?"

Talking it over with my Dutch friends, we examined every bit of evidence until, finally, Emmy Quant, another of my students, took a look at the *Avenue* issue offered to the woman who was now my avowed enemy, Emmy looked at the first line of the story I had written and blushed.

"It was this," she said, pointing at that opening phrase.

"What's wrong with it?"

"It's a bit strong. In fact, it's extremely strong. Words that are not mentioned, let's say, in polite company."

Suddenly I understood. This story, "En Familia," was about a young army recruit in post-coup Chile who returns home to a jobless and proud father who abhors the military, and the first words out of his mouth are *Tenía que ser mi puta suerte* — something like, It had to be my fucking luck. When Tessa, the translator of the story, had brought it to me in Dutch, I had noticed that it opened with *Godverdomme*, It had to be my goddamn luck, removing the irony of having my protagonist speak about a *puta*, what his twin sister had become, prostituting herself to feed the family and mirroring the soldier's own fate. Because he had also been turned into a sort of whore, Pinochet's whore, brother and sister trapped in a

Chile they could not flee, desperately seeking to rescue some zone of tenderness that had not been tainted by the dictatorship. "Spice it up," I instructed Tessa. "Use the most vulgar sexual expression you can think of."

She dubiously spelled out the word for *cunt,* and I nodded enthusiastically, yes, that'd be perfect, a male macho invoking female genitalia, unaware that he was referring to his own sister.

"*Avenue* might not like it," Tessa said. "They may ask us to change that word."

Unfortunately, *Avenue* accepted my suggestion. I had screwed myself, to use a term that the Housing Authority woman would have rejected. My literature, my overkill instinct, had obviously offended her to the point that she had decided to remove this pervert from Amsterdam.

What made that threat more worrisome was that it came on the heels of an incident so insolently coincidental that I still have trouble believing it really happened.

One Saturday afternoon late in 1978, I was crossing the Muntplein in Amsterdam, with Angélica and her colossal belly in tow, when a large, luxurious car that had been idling on the crosswalk suddenly accelerated, ignoring the red light and lightly nudging Angélica. I turned and landed a solid kick on the miscreant vehicle, vicariously channeling a rage better reserved for faraway fascists. But Pinochet wasn't driving that car, whereas its owner, alas, was extremely close by. Possessed by a holy wrath—he was protecting his property, after all—that short, irascible, powerfully built man with silver hair verbally assaulted me and might have tried something more physical if it had not been for the crowd berating him for putting our lives in danger. One of the bystanders suggested that I lodge a complaint at the local police station, where I once again traded nasty looks with our new enemy, already there to level his own charges.

I did not give the matter further thought until I received, a month later, a letter from the assailant demanding that I pay for the damage to his car or face the consequences. An idle threat, I felt, until my friends informed me that *deurwaarder,* the man's profession, noted on his stationery, meant bailiff, one of those bastards who remove tenants from their homes and seize their belongings in order to pay for arrears.

All of a sudden, a Dickensian future flashed in front of my eyes. Come July, there would be a knock on the door and he would be there, our foe, ready to exact retribution. Now it is my turn, Meneer Dorfman, let's see who gets the last kick, let's see who can best protect his property now, Meneer Dorfman. And behind him I conjured up the image of her, the Medusa-like wraith from the Housing Authority: Who are you calling a cunt now, Meneer Dorfman? How do you like it when your own life turns into a literary nightmare, when your fucking luck runs out like it does for your fucking character, who is just as destitute in your story set in Chile as you will soon be in Amsterdam?

That feverish scene never, of course, came about.

Joaquín was born and everything was promptly resolved.

Or rather, to give credit where it is due: the godfathers and godmothers of Joaquín were not going to let the new Dutch baby or his parents be homeless again, not in their land. Emmy Quant, who had just taken over the coordination of the hundreds of Chile Committees in the Netherlands, finagled a meeting with the mayor of Amsterdam, to which all our good Dutch friends went, along with a host of irate supporters of the Chilean cause. A message was delivered: if the Dorfman family was not spared, Emmy and her mates would occupy the mayor's office and initiate a protest that would shake the rafters of the municipality.

Two days later, I received a letter stating that, given my special circumstances, we could remain in our apartment until August 1980. And soon after that we were informed by the police department that the driver who had almost run us over had been reprimanded for recklessness.

Jubilation! Again I was being saved, again I had gambled that everything would turn out glowingly, again I had been right. I should have seen these troubles as a sign, but I was blinded by how, at each critical moment in my life, I always somehow managed to avoid disaster. In spite of the coup and what it should have taught me, in spite of my many exiles and close calls, I pictured myself as a golden boy, perpetually able to talk his way out of trouble. I assumed, even after death had stalked me, perhaps because death had stalked me and let me go, that I would somehow land on my feet. This incorrigible optimism was as essential to my persona as my boundless energy, feeding my capacity to bounce back in the most adverse circumstances. That attitude, typical of many sons of immigrants and invigorated, in my case, by a buoyant, can-do-anything American childhood, nevertheless had a downside, led to a pattern of steadfast refusal to accept that fate could someday turn against me, that life is not a game, and if it is, then at some point you best be prepared to lose, be prepared to get shattered by destiny.

Exile still had some cards to play, some surprises in store.

Fragment from the Diary of My Return to Chile in 1990

AUGUST 23

Joaquín has been crying.

He sits at the dining room table with his homework in front of him and outside the winter day darkens, and inside it is dismally cold, it's been like this every school night for weeks, and my only inane consolation is that maybe this is good for him, maybe this will strengthen him for what, given his heritage and the shipwreck of this world, will be a difficult life ahead, maybe this is what he needs to go through to break the curse of the wandering Dorfman tribe and put an end to the many exiles that have plagued the family. But I don't tell him this, and he tries to choke back his tears, he smiles up at me as I stand over him and caress his hair, he does his best to pretend that it will be all right, because this sweetest of boys doesn't want his sorrow to spoil my return — oh, and I had fooled myself into believing that my heart could not be broken all over again.

Every morning, before dawn, I wake him in the freezing house. A late sleeper — an effect of the nightmares vexing him for the past three years — he can barely pry open his eyes and hastily get dressed in time for a private van to take him to school on the other side of Santiago, an hour's drive away, an expense incurred so he could make friends with other kids on the way there, that was our plan, but he hasn't made friends, not in that van, not at school. In fact, his one pal, a boy named Orlando, invited Joaquín to his birthday party early one morning and then a few hours later *dis*invited him, saying his father's abrupt illness had forced a cancellation. What a lame pretext! From a nearby table a group of Orlando's buddies cheered Joaquín's mortification; it was clear they'd encouraged this spiteful act. And everything else at this preppy American institution has been equally disastrous: Nido de Aguilas may preserve the English of its students, but it offers the same asphyxiating traditional Chilean education as the rest of this country.

A shocking contrast to what Joaquín left back in Durham. To spell it out, in case this diary ever sees the light of day and Chilean readers wonder what all the fuss is about, well, Joaquín's been attending a superb Friends (Quaker) school in the States and that's where his true friends are. They write to him, call him on the telephone, insist how much they miss him, while here his teacher sends us messages accusing our child of not working hard enough at his Spanish. My God, will these two languages now at peace inside me start to brawl by proxy in my younger son? I write back asking the teacher to understand how much he is trying to learn Spanish. Oh, I could list ten incidents like that one and probably more, if Joaquín would tell us, but he doesn't complain, he just cries when he can't help it, our son is crying and there's nothing I can do to make him stop.

And to think that when he was born in Holland, on the coldest night of the century, I did not doubt that we would be able to spare him trouble. More than that: I was convinced, and did not hesitate to ecstatically broadcast, that our new son had come into the world for the express messianic purpose of purging us and that world of those terrors.

The streets were speckled with deserted vehicles and so slippery with ice that we needed a police escort to help us reach the maternity ward, the Nicolaas Tulp Kliniek, on the outskirts of Amsterdam. We would never have made it on our own, Angélica and I and, more crucially, Joaquín, ready to come out and save the planet on that dawn of February 16, 1979.

In Chile, twelve years earlier, I had been forbidden to witness Rodrigo's birthing, peremptorily banished by hospital policy; now, in a banishment that was not a metaphor, in a frozen land far from my own, I ended up serving as the midwife or close to it, one more of the strange, paradoxical gifts of being uprooted, that one can break away from the stifling customs of a conservative fatherland and become a real father.

That night at the hospital, Angélica had been holding the child in for hours because the obstetrician was late: the alarmed Dutch nurse had at first forced Dr. Krammer to plow through the blizzard at two in the morning and he'd gone back home once he realized that delivery was not imminent. With every contraction bringing a new wave of pain and another stern warning from the nurse that the baby could not possibly arrive until the doctor did, I decided to play the lunatic, the zany foreigner on his most atrocious behavior. I threatened to begin to scream up and down the corridors, and at last she relented, though she still moved slowly to prepare Angélica for the final stages of labor, so that when Dr. Krammer careened through the door with his overcoat glistening with snow, he was astonished that she hadn't been transferred to a proper delivery room.

He took one look at Angélica and grabbed one of her legs and ordered me to hook my arm around the other one and told her to push in his Dutch-inflected English, push, Angélica, push with every muscle in your body, push, push, and I said *empuja, mi amor, puja, puja,* and the good doctor repeated my words in a Spanish he did not know, *puja, puja, Angélica!,* using that rasping guttural Dutch *ja* sound — I would have laughed if I hadn't been so affected. Because a minute later I saw the tip of a head edging out, the hair black like Angélica's, and then the ample forehead and eyes closed so tight and the wrinkled nose, the face, that face, that color, the baby looked like Angélica's dead father. I blurted it out in Spanish, *es igual a tu papá, Angelita, igual a Humberto, puja, puja, mi amor, empuja,* but she had no ears for my lyrical litanies or the entreaties of Dr. Krammer, who was speaking to her in Dutch, *pars, pars,* she took another deep breath and *parsed,* pushed, *empujó* with all her strength into one final slick moment of pain — she was splitting in two, she was roaring a gentle fury of words in who knows what antediluvian tongue of tongues — and he was out, he had arrived, Joaquín Emiliano Alonso had been propelled into this planet by three languages and was giving his first cry in the one primal language of hurt and elation that every living creature on this earth understands, and I was sobbing for joy, as if all the horrors and sorrows of the universe, all recrimination and all guilt, all had been swept away, emptied away, by this simple tiny crying miracle in my arms.

Our answer to the death around us, the death that had engulfed our lives.

A sign that everything would change, everything was already changing. Heralded by the innocence of our new son, his eternally sunny disposition in the decade ahead.

Not that he hasn't had his share of the strife. He must have absorbed something from the tension of the family getting stranded in the States and those jobless first years in Washington in the early eighties and no residence status or health care and soon enough our many destabilizing returns to Chile, even a perilous trip to visit the *relegados* in 1985, activists banished by Pinochet to remote islands in the south of the country, outwitting the police and the army, pretending we were tourists rather than messengers of hope. All those dangers threatening us and him, but all through the turbulence, we were careful to shield Joaquín, perhaps obsessively, as a way of compensating for not having been able to do so with Rodrigo, no lost rabbits in Joaquín's existence, no books stolen by comrades, no angry Parisian fingernails flailing in his face, and definitely no politically charged nightmares, not even during the fraught seven months in 1986 that we spent in Chile, our first prolonged return that was supposed to be, like this one, forever.

But we couldn't. We couldn't shield him from hell, not him and not ourselves and not another boy about to cross our path.

And now Joaquín is crying, and there is nothing I can do to stop his grief now that he is living his own exile, repeating his father's fate, his brother's fate, his grandfather's fate, so many Dorfmans losing their land because of circumstances beyond their control.

Pray, that's what.

All I can do is pray that Chile will enthrall him as it did me, that the country I fell in love with has not died.

•

WHAT INTERVENED in Joaquín's life and our lives with such drastic finality, what changed everything for him and for us, was somebody else's son, another Rodrigo — not ours, thankfully — but a Rodrigo we knew and tried to save and could not.

Rodrigo Rojas was a troubled, bright, soft-spoken kid living in Washington, D.C., with his exiled mother, Verónica de Negri. We'd befriended the young Rojas boy on arriving in the States in 1980, kept up a relationship as the years went by, a lost soul, we thought, as we watched the child grow into a melancholy, pensive adolescent who'd found his calling in photography and as a budding computer wiz. He seemed troubled by the torture his mother had endured, savage gang rapes in a detention center in Valparaíso after the coup, didn't know how to handle his feelings of resentment against her. When other exiles in Washington, like the Dorfmans, started to trickle home and Verónica was still forbidden from returning, her son decided to head for Chile in early 1986, hoping to uncover, in the labyrinth that was his country, some clue to his own identity, perhaps meet up with the father he had never known. It was a rite of passage similar to what our own Rodrigo, also nineteen years old, was going through at the time, preferring to spend time in Chile with a semi-clandestine film group recording the struggle rather than rushing off to the feral pleasures of an American college. As for Verónica's boy, my intuition was that he couldn't leave his childhood behind until he had found out where he came from. Was he American? Chilean? Neither? Both?

I hadn't seen Rodrigo Rojas in Chile until, near the end of June 1986, I happened by chance to spot him at an opposition rally in Santiago, snapping photographs. Before he'd gone off to join a bunch of boisterous friends, I handed him the address of the Zapiola bungalow to which we'd just moved, sensing that he had something important to tell me. A few days later our young friend knocked on our door, precisely the evening when we were holding a small housewarming dinner. He peeked in at that group of adults sipping pisco sours and excused himself, turned to leave without stepping across the threshold. He and I covered our embarrassment by promising to see each other before I traveled to the States, maybe he could bring by something for his mom.

He never got to tell me what was troubling him, what he needed so urgently to confide to me.

One week later, as I was helping Angélica pack for our upcoming trip back to the States, I heard on the radio that a couple of kids had been found that morning, burned half to death on the outskirts of Santiago. I only half heard their names. Horror had become so normal. We would read that somebody had died in police custody and add one more name to the endless inventory in our heads. Three people had already been shot dead in the midst of a national strike that was paralyzing the country, one of them a thirteen-year-old girl who had gone out to buy bread.

And now they were burning youngsters as well—a new technique in terror, when would it ever stop? And then I went back to selecting documents I'd need in the next few months for a book I was planning, mostly notes from hours spent with psychologists and psychiatrists who treated the victims of violence while coping—and here was the literary hook, the perverse twist—with a similar violence in their own threatened lives, therapists and patients in the same sinking boat. One member of the group, Elizabeth Lira, had become close to me. She had emphasized how necessary it was to keep your distance from the repression, how necessary for your sanity, how necessary and also impossible.

As I was to find out when the telephone rang an hour later and learned that Rodrigo Rojas was one of the two injured youngsters. Sixty-two percent of his body had been severely burned, it was unlikely he would survive, and could I help through my contacts to persuade the government to permit Verónica to come to Chile, find a way of relocating the boy to the Hospital del Trabajador and its burn unit, among the best in Latin America?

Given the ongoing strike and the insurrectionary mood of the country, it was a bad night to play the Good Samaritan. I called the U.S. embassy and the Catholic archdiocese and anybody else I could think of, applied pressure so a distraught Verónica could catch a plane that very night, feverishly wrote down names and numbers in the shadows, because the electrical lines in Santiago had been sabotaged, shouted into the telephone, trying to negotiate with the officer standing guard outside the victim's room who refused to let the boy be moved. And all this over the banging of pots and pans and trash cans, the whistles and chants and staccato of faraway machine-gun fire and homemade bombs, as I scribbled information while outside the horizon flickered, lit up by hundreds of barricades and the paler glow of thousands of candles lining the streets to commemorate the dead.

Started, as well, to piece together what had happened to Rodrigo Rojas and to eighteen-year-old Carmen Gloria Quintana, the girl who had been assaulted along with him, how an officer had, after consulting a superior, doused them with paraffin and set them on fire. The bodies were wrapped in a blanket and dumped at a site near the airport where, a year earlier, the corpses of three major dissident figures had been discovered with their throats slit. The choice of that ditch sent a message to anyone daring to rebel: this is what awaits protesters, this ditch again, this burning.

A message made grimmer when, two days later, I received the news in Washington that our young friend had died of his wounds, died while his mother was massaging the soles of his feet, because there was no other part of his body she could touch to say goodbye. I received that challenge from the dictatorship and chose, acting as deliberately as the officer who had killed that boy, to ignore the warning, to respond by throwing myself, heart and soul, into a campaign to seek justice for Rodrigo Rojas.

The rest of 1986 and a good part of 1987 were spent hammering Pinochet without remittance, first at a press conference in Washington in early July, organized by the Institute for Policy Studies. Dozens of reporters and cameras showed up, enticed perhaps by an article on my seven months in Chile just published in the *New York Times Magazine.* I contributed relentlessly to the ensuing whirlwind, with an op-ed in the *Washington Post,* then with a long, denunciatory essay in the *Village Voice,* followed by interviews and debates on *This Week with David Brinkley, Nightline,* and *All Things Considered;* confrontations with Senator Jesse Helms on two TV programs, where I rebuked him for stating that Rodrigo Rojas was a terrorist; and on and on, including a faceoff with the Chilean ambassador to the United States, Hernán Felipe Errázuriz, on Charlie Rose's CBS show—maybe that's what had done it, the evisceration of that particular opponent, my attempt to avenge Rodrigo Rojas and all the Rodrigos of Chile, that's what may have tipped the scales against me. Or maybe it was the op-ed in the *New York Times,* after Pinochet had escaped assassination in September 1986, in which, while condemning violence, I wondered why it had taken Chileans so long to lose their patience, and predicted more bloodshed if the dictator wasn't ousted peacefully.

At any rate, after so much activity (and there was that false report of my death in Chile), it made sense to defer our homecoming, until finally, reassured by the Chilean consulate's trouble-free renewal of my passport and by a cohort of friends and acquaintances ("The government won't touch you—they can't be that stupid"), I set out for Santiago in August 1987 to test the waters. I wanted to visit my father in Buenos Aires anyway, to celebrate his eightieth birthday that month. But going to Chile first seemed foolhardy to Angélica. "Nothing's going to happen," I assured her. "Everybody in Chile says the worst of the storm has blown over."

"Everybody in Chile is, as usual, wrong" was her retort. She was not going on this *aventura loca,* she said, not this time. She was staying in Durham, was going to rent twenty French films, and would join us in Buenos Aires a few weeks later. "You go ahead with the boys if you're so keen, if you're so sure nothing will happen."

My first minutes on Chilean soil that morning of August 2, 1987, dispelled any misapprehension. The bored functionary sitting in the glass immigration booth hardly looked menacing as he tapped my name into the computer with crisp efficiency, his mousy, clean-shaven face more typical for a bank clerk than a member of Investigaciones, Pinochet's law enforcement agency. With a yawn he checked Rodrigo and Joaquín, stamped our passports, and handed them back unsmilingly, though he did manage a wan *Bienvenidos a Chile.* Rodrigo, however, had an oblique view of the detective's screen and something made him frown.

"What? What's the matter?"

Nada, he said, shaking his head, it's nothing. I dismissed my own nervousness and concentrated on the pleasant customs officer with sideburns and florid cheeks, who asked me from behind a long, low inspection bench if I had visited any farms, did I have anything to declare?

My playful answer: "Only a steak sandwich."

The sandwich was in my carry-on. The night before, in Miami, Eastern Airlines canceled one of the two flights to Santiago. When the one overbooked plane was about to depart, the agent chose us over hundreds of beached passengers, handing me three boarding passes, upgrading us from coach to first class. It was a stroke of luck that seemed a harbinger of better things to come, another disaster turned into another small victory. And I was content to have special attention, as Joaquín, in fact, was feeling ill and vomited during the flight. A helpful stewardess, noticing that the boy had skipped his meal, prepared a delicious steak sandwich, Argentine beef in a baguette. Joaquín took two bites and kept the rest for later, for the moment when I hauled it out of my bag and handed it to the *aduanero*. He examined the sandwich with a droll expression, raised his eyebrows, and said he was sorry but the item couldn't come in: "The kid eats it now or he doesn't get to eat it at all."

I remember the scene as if it were happening right now.

"No way," Joaquín pipes up in English. "I wanna play with my cousins." He can see them jumping up and down less than twenty yards away behind a glass partition, next to aunts and uncles and his grandmother, Abuela Elba.

The customs official confiscates the sandwich, carefully rewraps it, and places it inside a drawer—oh, the rascal, he loves that beef, he'll have a feast, he may have an eager son—but that's fine, I'm back again in my country and all the intimations of catastrophe have proven false.

Then someone taps me on the shoulder. It's the agent who stamped our passports minutes ago, flanked by another man, bulkier, with heavy eyelids and a pasty face, no mustache.

"Would you please accompany us, sir?"

"Why?"

"Just come quietly, please."

We are led into an office behind the customs area.

"You don't have permission to enter the country, Mr. Dorfman." It is the larger of the two men who speaks, joined by several ominous-looking policemen. "Your name was misspelled, sir, confused with somebody else."

I insist that they check the records again. The detective listens courteously, and I understand that I am to be deported, not jailed or tortured or executed, which swells me with a self-assurance I am far from feeling. In some corner of my soul, I'm even relieved: finally I'm being repressed, finally I'm a real victim!

"You're making a big mistake. You don't know who I am. Call the Ministry of the Interior."

"It's Sunday, there's nobody there. And this is not an error.

Your name is on a list of undesirables, *indeseables,* who can't enter Chile. That's who you are."

"How come I didn't know I was on that list?"

He explains — this is getting bizarre — that there's a secret decree, published in an official newspaper that does not circulate publicly. I conjure up an incongruous tableau: Pinochet, all by himself in his palace, reading a paper that nobody else gets to peruse, issuing decrees one day and reading them to himself the next morning, the ultimate model of the solipsistic dictator. But this is no laughing matter, no time for fantasies. I hand my captor the home phone number of a man in charge of emergencies at the Ministerio del Interior — a contact culled from my forty-eight frenetic hours trying to save the life of Rodrigo Rojas.

The head detective is impressed. He has heard of this man, he knows that he is powerful, maybe my arrest is a gaffe. He makes the call right then and there, gets an answering machine. "He's away for the weekend, won't be back till Monday," the bulky detective says. "And by then, you'll be gone."

"What about my boys?"

"They can stay as long as they want."

Joaquín breaks down. He is eight years old and doesn't understand why he's not outside with his cousins, what these men are going to do to his dad, perhaps to him and his brother.

I take Joaquín in my arms and rock him gently.

"The little one remains with me," I say, and glance at Rodrigo, who nods. "My eldest son is twenty. He'll stay and visit the family. Can he go out, allay their fears, then come back for his bags?"

The man in charge nods. He seems as adrift as I am, neither of us having lived through anything quite like this. Concerned by my connections, the possibility of a reprimand, he'll make things easy for us.

I feel strangely calm, as if some double of mine has already lived this scene and I were watching him from some safe zone. Absurdly, I think of retrieving Joaquín's sandwich. If they won't admit us, we have the right to keep it. If it weren't for that damn sandwich, I'd have made it outside, hugged my loved ones, reached Ana María's house—I didn't write the address on my entry form, maybe it would've taken them days to find me—ah, the thrill of being undercover again, hunted again as in the days after the coup.

Rodrigo interrupts my meanderings. He's back from his quick sortie and we look at each other and, yes, we'll use the obvious fluster of the agents to our advantage. The bulky man gives Rodrigo permission to transfer the plethora of gifts for friends and family to his bags. Rodrigo opens the largest suitcase and, concealed by the lifted lid, he snaps a photograph of Joaquín sitting on my lap, desolately fondling a stuffed toy dog, baptized Fernanda. I can't believe my elder son is this reckless, I can't believe that I'm delighted at his irresponsibility and courage, that I'm so fixated on getting this message out that I have not made him desist from an act that could spell his doom. But our jailers are doing their best to respect our privacy, the clothes are flying from one bag to the other, all three of us burrowing into trousers, toys, packages, like something out of a Marx Brothers film, a hint of how surreal things will keep growing.

Rodrigo leaves us but pops into the customs office several times as the afternoon wears on—for some reason, his first care-free crossing has been taken as a sort of safe conduct—and he uses these occasions to convey updates on the scandal that is beginning to rock the world outside. He has talked to Angélica on the phone, ruined her viewing of an Eric Rohmer flick. She has already called the *New York Times*, the *Washington Post*, the *Los Angeles Times*, the London *Guardian*, *Le Monde*, the major TV networks. I have suggested, via Rodrigo, that she contact every outlet I'm associated with, and imply that my arrest is due to my specific relationship to that media organization. Meanwhile, I've been engaged in an abrasive discussion with the men who arrested me. They want to use my own return ticket to deport me, and I insist that if the government wants to expel us, they should pay for it. Finally, there is nothing to be done except nurse lunatic plans to sue Eastern Airlines for agreeing to participate in this human rights violation.

Around this time, the head of airport security permits Joaquín to accompany his brother to fleetingly hug the family, all of them still waiting. And when Rodrigo returns with Joaquín, he brings along my nephew and two nieces — it's a hurried hello, a protracted goodbye to the three of them, and they break down my emotional defenses. I'm affected by their innocence, how they clamber onto Tío Ariel's lap and ask for a story, ask why can't I go home with them, Amparo and Matilde and Matías. Once they are shepherded away, Joaquín and I are left in this stuffy room whose only furnishings are a bedraggled pair of chairs and an ancient typewriter under lock and key. The two men in charge of my imprisonment offer to buy coffee, fill me in on details of a detective's life for a possible novel I'm planning, get me food which they won't let me pay for, even propose a short walk on the runway to catch a glimpse of my beloved mountains. One of them asks, when the other is not listening, could I autograph a volume of my short stories he has at home — if I ever come back, that is? And the other turns out to be a chess player, and we sit down to a game. He's not very good and I contemplate whether to let him win, and then decide what the hell, this is the only victory I'm going to be granted today and so I smash him to bits, check and check and checkmate. After a second game, he gives up and starts to play checkers with Joaquín, and they're still hopping and jumping their pieces when the news comes that our plane is ready to depart.

As I bid farewell to our captors, I realize that I've spent a whole day with the very men who haunt my fiction, yet I am no nearer now than before to solving the mystery of what or who they are and why they repress. How to explain the excessive politeness of people who have witnessed and probably perpetrated who knows what crimes? Traditional Chilean hospitality? The awe that many lower-middle-class Chileans feel for those who seem important or well connected? Would they have been as deferential with a coal miner, a barefoot peasant? Were they trying to communicate to me that they are disaffected with Pinochet? Or is the episode just one more example of the schizophrenia that accompanies the institutionalization of a dictatorship in a country that used to be democratic and where once in a while you uncover, in the midst of the most brutal confrontations, islands of decency?

Idle speculations. These men have always done their jobs. Thousands of former exiles now living in Chile are wondering, at this very moment, if my case is a portent, if they too will be expelled again.

I am hustled onto the plane, to the astonishment of the same stewardesses who took such good care of us in first class just last night, the very woman who prepared Joaquín's steak sandwich gapes as we are hurried to our seats — in coach, of course — by the four plainclothesmen. My face is red with mortification as every eye in the cabin trails us. They must think I'm a drug dealer, a child abuser, a terrorist.

As the plane guns its engines, I clutch Joaquín's hand, awaiting the moment when I will once again no longer be in physical contact with Chilean territory, and the tension, not only of the day, but of the many years of fighting the demons of distance, catches up with me. I feel weary, deeply sick, as if my bones wanted to vomit; tyranny works in dizzy, circular ways, repeating itself until you're ready to scream, enough, I give up.

Fourteen years before, on my first banishment from Chile, this was the exact route I had taken, sped down this identical runway. Then I had been able to comfort myself with the lie that I would be back soon. Ten years passed before I saw the mountains again.

How long would it be this time?

It turned out not to be long at all.

The international scandal was so immense, the disarray in the conservative coalition backing Pinochet so severe, the news coverage so hostile, that two weeks later Joaquín and I returned triumphantly to Chile, accompanied by Angélica, who had come to meet us in Buenos Aires. A plethora of reporters and photographers awaited us on the tarmac, microphones thrust in my face, TV cameras whirring. The customs agent who had stopped me now greeted us elatedly, swearing he never ate that sandwich, next time he'll let me enter the country with as much food as I want. And when Joaquín spied the plainclothesman who had played checkers with him, he hailed him with naïve familiarity in heavily accented Spanish, "Hey, when are we going to finish our game?" The response: "I hope we don't need to."

Adding to the lunacy was the news from Arturo Navarro, the publisher of the first edition of *Widows* to appear in Chile, after it had been rejected by the original publisher as too dangerous to bring out, even under a pseudonym. Thanks to Pinochet's attack on me, the novel, under my very own name, has become a bestseller. And the next day, at a book signing, things got more peculiar. After the stock of books sold out and the overflow crowd started to leave, I was approached by two men in dark suits — not your typical Chilean intellectuals. Were they going to arrest me again? No, all they wanted was my autograph on their copies of *Widows*. No need to inscribe it to anybody in particular, but if you wouldn't mind, Señor Dorfman, please write the date and the exact time, yes, there, thanks.

Brazen to the point of foolishness, I asked them why.

"In case we have to prove we were really here, sir," the elder one said to me. "We wouldn't want our boss to think we're shirking on the job."

Not all was quite that humorous during that return. My euphoric mood at having put the government on the defensive began to diminish as soon as Angélica and I were forced to contemplate the long-term costs of that whole incident.

Joaquín had confessed on the plane to Buenos Aires that he felt like a "tangled wire," and that very night he awoke sobbing from a nightmare. I managed to comfort him, show him that no monster was lurking under the bed, tucked him in, promised I would not leave his side. That was when he asked if his mother had been killed—not an illogical conclusion, given her absence. Those dreams of blood and slaughter, that recurrent fear of wild things hiding in the shadows of every room, did not disappear, nevertheless, when Angélica arrived in Argentina.

According to my psychologist friend Elizabeth Lira, who has treated a multitude of traumatized victims, some younger than Joaquín, that reaction wasn't surprising. Sheltered as our happy-go-lucky son might have been, a sponge inside his mind must have soaked up all the talk about terror, the casual conversation, no matter how hushed, about Rodrigo Rojas burned alive and José Manuel Parada's throat being slit and the reminiscing with Pepe Zalaquett about his time in a concentration camp, the steady, grinding, everyday drumbeat of repression surrounding our family, all of which had presumably been consigned initially to some safe archipelago of fantastical events, akin to what Joaquín had read in fairy tales, no more menacing than invented ogres and sorcerers. And at precisely the age when children begin to distinguish between what is imaginary and what is real, the floodgates of undeniable, historical violence had opened up and swamped him. It was the time of the demons—every one of them so effortlessly dismissed in the past was being validated and could now come out of the darkness to devour him. The father who was supposed to protect him, a bulwark against the beasts of the mind, had been himself taken into custody by the beasts of history; the defenseless father unable to stand up for the vulnerable child won't be able to defend him from another onslaught tomorrow, cannot be believed when he says it was just a dream, don't worry. It will take a long time, Elizabeth cautioned us, for this child to recover.

How long?

I didn't ask her, hardly dared ask it of myself, beset as I was by the old pangs of guilt. I had rushed forward without fully weighing where my actions would lead, hoping that everything would turn out right, blindly trusting in my good luck. It was because of me that our boy and his soaring tender spirit had been poisoned, because I was unwilling to give up the struggle against Pinochet, did not succumb to silence. Or was I to let Rodrigo Rojas be torched and his murderers get away with it, could I have acted differently? How could I live with myself if I gave up so my own child would be safe?

In every man's life, in every woman's life, in every family's life, a moment comes when more prudence is called for, and that moment came for me, came for us, once we returned to the States from our misadventure.

Perhaps if we had not been uprooted over and over during these years, wandering from Argentina to Havana and then Paris and then Amsterdam and the cats and the steamer across the Atlantic and the uncertainty of shipwrecked years in Washington and then our move to Durham. Perhaps if we had endured only one or two cities, perhaps it would have been possible to have risked the unstable mess that was Chile under the dictatorship. When we went back to Santiago for the first time, in 1983, we envisaged possible imprisonment, beatings, torture, persecution, even death, but not the more remote possibility of a new banishment, not that moment of dizziness when I found myself again on the plane heading for Argentina as I had done after the coup.

It was as if the earth beneath our feet had yawned open, our lives made again of quicksand. There was no guarantee that next year I wouldn't be expelled once more — and if that were to happen, there would be, I knew, no stamina left to roam the earth in search of a country and a visa, to start all over from scratch. We had a green card that allowed us to remain in the United States. I had a position and a salary at Duke, marvelous colleagues, eager students, and health coverage. We had a school waiting for Joaquín, and Rodrigo enrolled in UC Berkeley, and a house in North Carolina, a house with our furniture and books, and a publisher and editors and friends like Mark and Susan Schneider to protect us, and security, security, security. This is the truth: it would have been negligent to abandon that American sanctuary it had taken us so long to conquer.

And so, once we were able to assess how extensively that traumatic experience marked us, we chose not to return permanently to Santiago until democracy had also returned. We added three more years to our exile, three years spent away instead of living through the process of transition in Chile, three lost years that were to prove decisive for our final break with the country.

Such a minor form of repression that we suffered after all, that day at the airport, that jolt of understanding that we were to be deported, those plainclothes policemen who did not leave so much as a bruise on our bodies. Such a small intervention in our lives, and yet we continue to bear the scars.

Joaquín, for all his success now at the age of thirty-one — a consummate writer, three novels published by Random House, each translated into several languages — is still paying the consequences of that shock. Just as Angélica and Rodrigo remember, in the deepest recesses of their minds, if not on the everyday skin of their lives, how our existence was twisted by men we had never met.

In this, my family is not exceptional. I can transmit the warped details of their odyssey because they are closest to me, they are the ones whose pain I have been forced to witness, whose pain I cannot escape. But I'm aware that my family's troubles can give only a pale measure of the vast Chile damaged by the dictatorship, and maybe that's why I keep finding it impossible to tell my own tale without offering refuge to other stories, why the agony of others trapped in Pinochet's perverse legacy keeps intruding in this saga. Maybe that's why victims seek me out, recount their woes, won't allow me to stay within the confines of my own distress.

More often than not, these stories of affliction involve women. Two of these women awaited me with their trauma on that 2006 shoot. One I had been searching for over the years, and as for the other, I had forgotten her existence and remembered her only after she sought me out.

Adelaida was a baby when I had taken her in my arms, probably no more than a few months old when I'd cradled her in the Argentine embassy in Santiago one morning in early November of 1973 and tried to calm her cries with a Mozart aria. It was no use. The little girl was inconsolable, shouldn't have been there, had basically been kidnapped by her father, Sergio Leiva, who didn't have the slightest clue about how to care for her.

Leiva was a troublemaker. In a Chile brimming over with irresponsible, extremist left-wing splinter groups that could spout the glories of armed struggle as the solution to all of our ills because they knew that Allende's government would not repress them, his minuscule radical faction had been particularly immature and aggressive, stocking up on weapons and calling for the workers to create a dictatorship of the proletariat. And when the military unleashed its full brutality on the defenseless patriots of Chile, Leiva had not exactly stood his ground — as Allende had — but ended up (with his newest girlfriend) in the embassy, where he continued to espouse rhetorical niceties about blood and revolution.

In spite of all this, we struck up a relationship. The moody songs he composed and his fierce romantic streak appealed to me, and so did his hatred of injustice, no matter how misguided. And who could deny his heartache once it became clear that the baby had to be returned to her mother, who among the refugees did not feel a twinge of solidarity with him the day he had to say goodbye to Adelaida?

The girl's absence unhinged Leiva — not that he was very stable to begin with. He would provoke the police guarding the embassy grounds, give the finger to the sharpshooters who watched through binoculars our desultory lives from nearby high-rises, rifles ready. We kept asking him not to hurl any more profanities at them, and he would answer by accusing his critics of cowardice, of not being eager to continue the struggle. He became angrier and more depressed after the dictatorship refused him safe conduct to leave Chile. The last vision I have of him is from the day I left the embassy to take a plane to Argentina: bidding me farewell with his fist in the air as if he were Che Guevara, singing of a socialist future that was receding evermore into the mists of tomorrow.

I can't say I was entirely surprised when one day in early January of 1974 I read in Buenos Aires that Sergio Leiva had been assassinated. He had climbed a tree in the embassy's garden and had been busy insulting the junta and capitalism when a policeman shot him down. The dictatorship denied that version, insisting that Leiva had been killed outside the grounds while trying to jump over a back wall in search of asylum—ridiculous, given that he'd been a refugee for months. The scandal caused by this violation of diplomatic immunity rapidly died down, and Leiva's name was added to the long list of expendable victims, obliterated by history, and by me as well, only retained in some vague garret of my memory where atrocities accumulate like geological layers.

By everyone, that is, except Adelaida.

That daughter, who had hardly known her father, decided many years later—in fact exactly on Pinochet's ninetieth birthday, November 25, 2005—to accuse the former dictator of homicide, indicting him along with Ismael Martínez, the *carabinero* who had shot Sergio Leiva. I wouldn't have heard about this—it barely made the papers—if I hadn't received an e-mail, almost a year later. It was Adelaida. She had heard that I had been in the embassy and wondered whether I had met her father, the singer Sergio Leiva. Nobody, she said, was willing to testify that he had lived there during those last months of 1973. Would I help her bring her father's murderers to justice?

I answered that I could testify next month, as I'd be in Chile to film a documentary. And would she mind if cameras accompanied me on the day I appeared in court? Although it makes me uneasy when someone else's pain is exposed and filmed, I know from experience that victims seldom have such reservations: they want their story to reach the widest possible audience.

Adelaida was no exception. She was grateful for the opportunity, had so much grief to pour out. Hers was an existence devastated, she said, by the bullet that had killed her father. She broke down in the middle of our conversation, wept at the injustice of the world and her own fate, and then comforted herself by asking her small son to hug her, hug her very hard, very tight, very long. He was a mischievous imp of a boy who had inherited his grandfather's musical inclinations but, auspiciously, none of Sergio's wild political ideas. Already, though, he was bearing the weight of the family's history, caught in a whirlwind that had begun decades before his birth and might well spread to his children, one they would transmit like a virus to who knows how many generations.

I am not sure if my brief intervention in her life really served any purpose, stanched the pain. She said it was important, and more so when I managed to get her into the Argentine embassy—which she had not seen since that remote day in 1973 when she had been retrieved by her mother. She said it mattered to sit with her child under the tree where her father had been murdered, walk the grounds where she had been consoled as a baby, when tears of a different sort had streaked down her face.

"I want people to know," Adelaida said to me. "I want everybody to know that these things just can't be, they just can't be tolerated."

And yet what if the story can't really be told, fully told? What if the injured party prefers not to speak?

Because there was another woman I had met in the days after the military takeover and also not seen in thirty-three years, one more story of hidden damage I carry inside, more unfinished business I was trying to deal with on that 2006 trip.

This is not, however, about how I helped someone to recover the memory of her father. It is about how someone helped me to stay alive.

It was the end of 1973, and I was running for my life.

That's when Patricia had come to the rescue.

She drove up to where I had been hiding, in the chalet of a dip-lomat, one of the many houses where I had sought sanctuary after the coup. I knew nothing about her, only that she was part of a clandestine network dedicated to saving the lives of Allendistas, only that she had found somebody willing to secretly give me refuge, only that if we were caught we'd both be killed.

We silently crossed the city infected with soldiers and guns, she deposited me at a modest residence in the working-class suburb of Maipú, acknowledged my thanks with a nod and a half-smile, and was gone. The less we spoke, the less I could reveal if the soldiers caught me.

So there was a mystery back there in my past, and when we decided to film the documentary, *A Promise to the Dead,* one of my objectives was to track down and truly thank the woman who had been part of a vast and gentle conspiracy, the many patriots who had risked their own lives to save mine.

If the existence of baby Adelaida had vanished from my mind, my thoughts had turned now and then to the silent woman whom I now call Patricia during the seventeen years of exile, wondering if she had been spared the fate that afflicted the *desaparecidos,* asking myself if there was any way to discover the identity of someone who had tried so hard to hide from me and from the soldiers who she might be.

It turned out that she was safe and she was sane. Carlos Salas, the man who had given me shelter in that working-class neighborhood, had stayed in touch with her and managed to connect us. As she drove me down the same avenues from long ago, retracing our route, I learned her real name and her fascinating story.

The day of the coup, her estranged husband, an abusive right-wing thug, had returned to the house he had abandoned the year before. Licensed now by the military takeover to indulge in the worst excesses of his already sadistic personality, the man used

Patricia's home as a headquarters for a small group of fascist miscreants who captured and tormented Allende supporters. Like so many women placed in an impossible situation, she had kept her wits about her and used his presence as a cover for her own revolutionary activities. The car with which she transported me belonged, in fact, to her former husband, the very car in which he cruised the sinister city. That Pinochetista car had been my salvation.

And yet that extraordinary story, that name, that woman, are not in the documentary.

The streets of Santiago were no longer filled with soldiers, but the old fear still lingered in the air. She didn't want to be filmed, Patricia said, because right-wing members of her family — one of her sons, one of her daughters — hadn't the slightest inkling of her secret heroism, how she had risked everything to save people like me. If her identity surfaced on a screen, she added, there would be drastic penalties to pay. Chile in 2006, sixteen years after Pinochet had left power, was still contaminated, people like Patricia were still in hiding.

This was not how I had pictured our glorious reunion. Somewhat naïvely, I had assumed that, just as she had offered me redemption from death, the documentary crew trailing me around Chile would redeem her from an undeserved oblivion.

But if the camera inhibited one imagined scene, that same camera facilitated something else, perhaps more significant.

Two years later, I went to Chile to present the documentary at a cultural center recently built below the presidential palace, under the very building where I should have died and did not, an experience made more unforgettable when I invited all those who had been filmed in it (even those who had ended up on the cutting-room floor) to join me after the screening, all those who had kept me alive during the coup and in the decades that followed, one more surreal moment in my overly real life.

At the end of my acknowledgments, once I had brought everyone up on the stage, I called out the name of Patricia. With her authorization. She had agreed to step into the momentary spotlight. The time had come, she said, to face the truth of who she was, to have her children accept her full story, and too bad if there were consequences.

Why did you decide to go public? I asked her at dinner that night.

She had been afflicted by cancer, she said, when we had met in 2006, when she had asked that her story be suppressed, but that encounter—the second one of our lives—had worked some sort of miracle inside her, had restored to Patricia the deepest courageous core of her identity, and the result had been an inexplicable remission of her disease. She believed that telling me the story, even if it was offscreen, just getting it out of her system, being confronted all those decades later with one of the men she had saved from death, had helped her face her own death, had given significance to that life of hers, had helped some angel in her body to keep fighting.

So that's how life is. You take a child in your arms and sing to her and many years later the child is a woman in need of help. And a woman you do not know saves you from a firing squad and you give thanks by keeping her company in an hour of need. And a young man is burned to death and his mother says goodbye in a hospital room. And my Joaquín is battered by what his father was forced to do to keep stories like those alive, and must find his own way of surviving, that's how life is, that's how our lives turned out.

Fragment from the Diary of My Return to Chile in 1990

AUGUST 25

This morning I went for a long walk with Pepe Zalaquett up into the hills above our home in La Reina. Saturdays can be nice, there's much less traffic, smog, and congestion, and today, as I breathed in the pristine air, I felt grateful that Pepe, the head lawyer of the Comisión de Verdad y Reconciliación and, in many senses, its guiding light, was able to take a few hours off to catch up with me, seeing how busy he's been. The Truth and Reconciliation Commission is swamped with more witnesses than expected and new clandestine sites where bodies pop up out of the ground, and each case of a *Desaparecido* has to be documented, the commission's account of the circumstances under which people were executed has to be airtight — even if no trials are planned and no perpetrators will ever be accused. Fortunately, President Aylwin has not skimped on resources, and Pepe has assembled a staff of young lawyers devoted to this labor of love.

Though somewhat skeptical of the commission's limitations, I'm impressed by its passionate commitment and moved by my friend's portrayal of the peace of mind and heart being offered to the widows and orphans of the dictatorship. Beaten by the police, mocked by the authorities, and spurned by corrupt judges, they have at last been given the chance to address a governmental organization. "It's the first time in fifteen years, sir," one woman said to Pepe, "that I have even been asked to take a seat. They always kept me standing, each time we asked for information."

Her pain and the pain of so many others are being granted an official space. This sanction by the state makes that pain a legitimate component of History, dragging those violations out of the realm of conjecture and into the light of day, to be taught in schools, certified in speeches, memorialized in monuments, a Truth that can never again be disowned. Aylwin avoided setting up a commission composed only of opponents of Pinochet precisely because that would have made its conclusions acceptable only to people "on our side," preserving the divide of the previous decades. Instead, half the commissioners are conservatives, sympathizers of the former regime, albeit ready to form independent judgments. Decent people who can be persuaded by evidence. In fact, Pepe says, it is often these members who are most appalled at the horrors of the dictatorship, most adamant about the need to uncover every last repulsive act.

When I asked Pepe why the proceedings weren't being televised so they could be discussed on every corner and at every dinner table and at work and in the bars and throughout the universities, his answer made sense: what each victim states has to be corroborated and investigated, their pain and intimacy deserving respect. As for the commissioners themselves, the confidentiality of their deliberations guarantees that there won't be misunderstandings as they work out their doubts and reach a painstaking consensus. He doesn't say so, but I suspect that they are trying to prevent turning this tragedy into a spectacle, a media circus. Pepe has always been rigorous and discreet — very different from his exhibitionist friend Ariel, and probably for the best.

I didn't have time to pursue other questions. Perhaps I averted them on purpose, because they are awkward and I love Pepe and he's doing such an extraordinary job. But I'm worried about the victims who survived and have no place on the *comisión,* in the Official History of Chile, what about their distress? What about the women who were raped and the men who were jailed, the hundreds of thousands who were bludgeoned and rounded up, the kids who saw their fathers wrenched from their homes in the middle of a winter night, stripped of their last piece of underwear, and penned in the mud of a soccer field until dawn? What of Cecilia who can't look at that father and not see his shivering, primeval body in the rain? What about the exiles who left because they had to save their lives, and the migrants who left because they had no way of making a living? What about Rosa who has returned to a country where she can't find employment? What of a woman I met, Maribel, who went all the way to Valdivia where her boy was being held and who told the guard to let her in to see him or she would kill someone, kill the first passerby, so they'd put her in jail and she could protect her child from being raped? What about José María who survived a mock firing squad and still wets himself whenever he hears a car misfire? What about him, about her, about them?

I know it's impractical for a commission to receive all those people, to hold hearing upon hearing so they can recite their tales of woe, obtain some form of recognition or reparation—it would turn the transition into a wailing wall of laments, stagnate us in an eternal victimhood of regret, destabilize our democracy by repeating the divisions and recriminations that led to this tragedy in the first place. The problem is that such a line of reasoning, legitimate and judicious as it may be, lets the perpetrators off the hook, enables them to look their children in the eye each morning, allows those who gave the orders to anonymously and flagrantly roam the streets and sip their martinis. This fear of the wounds of the past leaves those who benefited and prospered from that terror to hold their heads up high, creates a haunted country where those who suffered and those who inflicted the suffering have to live side by side, everybody complicit in avoiding the truth of the irreparable harm that was done. That's what concerns me, what that evasion of the truth can do to us as a nation. Of course it's important to reach a consensus among former enemies or we will never advance towards a better tomorrow; it is crucial to deal responsibly with the human rights issues of yesterday so we can tackle today a host of pressing problems in health, education, environmental degradation, the inheritance from Pinochet's misrule; and it may well be that there are no conditions now, in 1990, for the stark sword of justice to be wielded. A few years from now the tensions of the transition may have mitigated and the army put on a leash, and then trials can begin and punishments be meted out for the most egregious crimes. I realize that impatience can be dangerous, and yet I am deeply troubled.

The relatives of the *Desaparecidos* told me, when I visited them last month, that they had lobbied President Aylwin for the commission to be named "de Verdad y Justicia," so that Justice would have as much weight as Truth in its findings. Aylwin finally chose to substitute reconciliation for justice, and though I can appreciate the pragmatic reasons for such a change in emphasis, I doubt we can engage in any genuine reconciliation until there is some measure of justice and our enemies have at least recognized the pain they inflicted and swear to never again repeat such actions. Otherwise, the phrase *nunca más* is merely rhetorical.

I write these misgivings hesitantly, with trepidation, because I've been wrestling with the dilemma of how you coexist with those you hate since before the military takeover, since before I was really hurt, before I learned how hard it is to live side by side with men who can kill those I loved, who could kill them and did, who could exile me and mine and did, who could arrest my little boy and did arrest him. This quandary has marked my life and is now central to the sort of nation we desire and deserve.

If I were to locate the moment when I first became aware of this ethical swamp that spills out of the need to accept the ineradicable existence of evil and yet struggle against it, if I can trace back to one incident the start of that discussion inside myself, it is to an evening in the Chilean winter of 1973, perhaps a month before the coup. Even though Allende's power was waning and our future looked dismal, I was still an inveterate lover of chamber music. So I decided to buy a ticket when I read about a concert in the Teatro Oriente. I would squeeze in the performance between an afternoon at La Moneda trying to imagine media alternatives to the right-wing offensive and a late-night session of painting walls with my buddies, a way of wagering that there continued to be a place for beauty as the Chilean *Titanic* sunk, that those fiddlers had been right to keep playing on the deck of a ship about to plummet into the depths.

But there was no solace from the conflict rampaging through Santiago; it followed me into the hushed concert hall, it sat down in the row in front of me, just to the side but within touching and spitting distance.

The man's name was Jaime Guzmán, and after the military takeover he would be Pinochet's closest adviser and draft his constitution, and now, as I write these words, he is a senator, the leader of the opposition to President Aylwin, the founder of the neofascist UDI, a man who has justified Pinochet's abuses as imperative to save Chile. At the time of that 1973 concert, however, he was notorious for his appearances on the most-watched television program in the country, where every Sunday he would spatter his zealous, Opus Dei venom with icy cold logic and strictly parsed syllables that had been practiced, rumor had it, while kneeling in front of a crucifix for hours, wearing a cilice and lacerating himself with a whip. And he looked the part: thin as a ghoul, sunken eyes peering from behind glinting glasses, with a head that was the caricature of a skull, anal-retentive and repressed, oh I couldn't bear him.

That was the man who, as the musicians began to play the first strains of Beethoven's String Quartet No. 15, the pinnacle of chamber music, an intimation of the divine even for this agnostic—it was that selfsame Jaime Guzmán plotting his murderous coup breathing close by me. More than breathing, I could hear him sigh with fulfillment at the notes, and I could scrape no pleasure out of this angelic music, no interlude of serenity, I had to strangle that man, or maybe just give him a smack on that comic-book skull, either that or leave, and I stood up and fled, back into streets where the battle for a world where my Beethoven would not be soiled by that man's moans.

It had been the easy solution, a feasible one at any rate, a preparation for my exile, in fact.

But now I am back here in Chile for good and must pass enemies like him every day on every sidewalk, encounters made more difficult by my knowledge of what he and his kind are capable of, now that I must learn to coexist with the ominous men who burned adolescent bodies and made wives into widows and created the sort of sorrow that Beethoven was born to assuage, that I was born, it seems, to confront.

•

SO THERE it is, the remote origins of *Death and the Maiden,* the idea of music joining deadly adversaries planted as a seed during that concert, but only blossoming into my life as a memory seventeen years later, when I first wrote it down in my 1990 diary. I had no way of knowing back then, on that now faraway autumn day, that Jaime Guzmán would be vilely assassinated by an ultra-left-wing commando in April of 1991, an act I repudiate, an act that had serious consequences for our transition to democracy, offering Chilean conservatives a martyr that they tried, and still try, to equate with the innumerable victims of the dictatorship's state violence. Today I would not describe Jaime Guzmán with the same hateful words, but I have left the extract as it was written, because it provides a clue to what was going on in my head at that critical juncture in my life, how I was already mulling over—though not yet ready to admit—some sort of literary approach to this dilemma, already foretelling the story of a woman who has been tortured and identifies, or thinks she identifies, the man who tormented her through the music they shared, as I shared that quartet with Jaime Guzmán.

This impulse to probe the relationship between enemies and music has a long genealogy in my existence; it did not suddenly spring into being out of nowhere. The next stage of its evolution came one afternoon during my Parisian exile. I was reading a magazine sent from Santiago by my ever-devoted mother-in-law, Elba, an interview with General Gustavo Leigh where he happened to mention how much he adored Beethoven's late quartets.

I threw the offending journal on the floor, had to keep myself from stomping on it. *Qué le va a gustar Beethoven a este hijo de puta,* it couldn't be that this sonofabitch loved Beethoven, the late quartets no less, no way could the general who had planned the coup and operated the ferocious intelligence service of the air force, no way could that devil love Beethoven. Many minutes passed as my overwrought heart slowed to a normal rhythm, and all the while I kept staring at that magazine at my feet, just Leigh and me and the stillness of that apartment so far from Chile as the evening darkened into solitude. And then I picked up the magazine, clicked on a light, and read the interview all over again. What if Leigh was indeed enchanted by Beethoven? Didn't that overlap make his position and mine more complex and challenging? Wasn't it the task of a writer — even if at that point I was still struck with silence and unable to articulate one noteworthy verse — to travel that divide, explore that juncture, that coming together of two sides that were glued to each other in ways that neither would want to acknowledge?

When I did recover my expressive skills, I began to tentatively delve into the intricate territory of the oppressors, take small steps towards puzzling out their identity. But it was only when I was in the midst of writing *Widows* in Holland that I had evolved enough, had looked into myself enough, to reach into the soul of a character — the soldier Emmanuel, an orderly who, in order to rise in the world, has betrayed his own impoverished family and will betray his lover — and to find it in me to pity him. For the first time I allowed myself to invite along on my journey somebody so wounded and damaged, to feel real compassion for him, a man I would not have wanted to meet in a dark alley or a naked basement.

If I was able to offer full humanity to my enemy, it was not only due to my returning literary skills. Something had happened in Amsterdam, something that made me drastically question myself. Without that momentous moral catastrophe, I would not be the person who wrote *Death and the Maiden*, the man who writes these words now.

It all started, as major crises often do, with an insignificant event.

One March morning in 1979, a month or so after Joaquín's birth, the telephone rang in our Buitenveldert apartment. It was Ankie Peypers asking if she could stop by. As our resident elder sister and permanent co-conspirator in all things Dutch, Chilean, and poetic, she would drop in at unexpected hours, but this time there was reticence in her voice, a holding back from her usual levity, from the cheerfulness she had exuded when she'd last visited, a week earlier, bringing a poem for Joaquín. She'd been so impressed by the deep wisdom in his newborn eyes, what they had seen from the womb or from some previous incarnation, that she'd written some verses—in Dutch, of course—which she recited softly to the baby, who looked straight into her eyes as she thanked him for teaching her how to begin.

You teach me to begin.

Words she might have been writing for me. Because Ankie sat herself down at the edge of an armchair, didn't accept Angélica's habitual offer of a glass of sherry, some crackers, asked me to swear that what she was about to reveal would remain between the three of us.

"Of course, Ankie. You know I can be trusted to keep my word."

A Dutch woman influential in both the solidarity movement and the political world of Holland had informed her that this Ariel Dorfman was working for the CIA. Given our friendship, Ankie felt obliged to alert me to this accusation, despite having promised not to do so. She had also been told that the money we'd been sending to Chilean writers, painters, and musicians had been pocketed by my party, that the MAPU political bosses were benefiting from the generosity of the Dutch and from cultural events organized across Europe.

I was more astonished by Ankie's tone of voice and body language than by the charges themselves: she kept shifting her eyes to avoid meeting mine, wouldn't sit back in the armchair, was perched on it as if about to fly off.

During the next two hours, she started to relax, agreed that the most striking indictment, that I could be at the service of the American agency that had helped to overthrow Allende, was incongruous, downright silly. The only proof of the truth of this allegation seemed to be my impeccable English!

I managed to coax from her the name of her informant. It was someone I will here call Sonia, a major force in the Dutch Salvador Allende Centrum, created by the mayor of Rotterdam. And where did this Sonia get her information? Ankie hesitated before revealing that Sonia was apparently "involved" with someone from Chile, one of the refugees, Jaime Moreno.

Moreno was a member of the Chilean Communist Party who headed Vrije Muziek, an organization vital to his party's finances and influence. Operating from Holland for the whole world, Vrije Muziek produced records, books, and posters, arranged mammoth concerts featuring Chilean musical groups—the Inti-Illimani, the Quilapayún, and Angel Parra, among others—taking in considerable amounts of cash and guaranteeing subsistence for muralists, writers, and filmmakers affiliated with the Communists, as long as they toed the party line.

There had been some tension with Moreno, signs that he and his superiors in the hierarchy were unhappy with my cultural initiatives and even unhappier with the philosophy behind them. The Centers for Chilean Culture (CCC) I had promoted were meant to make the artists themselves, inside and outside the country, independent of bureaucracies, free to be critical without having to fear their funds being cut off, a way of contesting the Stalinist idea that if you control the means of production, you hold sway over other forms of human activity. It was also a way to actively implicate foreign cultural figures in the solidarity movement — many of them stars and celebrities — to help their Chilean counterparts, to go beyond mere paternalism, open up a dialogue.

To the chagrin of Moreno, a number of Communist militants and fellow travelers had signed on to our projects. And, of course, money was flowing into Chile for cultural groups and individuals, and funds were being raised in Holland and elsewhere for events abroad that competed with those controlled by Vrije Muziek. To make matters more complicated for Moreno and his party's cultural apparatus, I had been named the cultural representative for the Resistance abroad by Allende's former minister of foreign affairs, Clodomiro Almeyda, who headed both the Chilean Socialist Party and the coalition of parties opposed to Pinochet, and this gave me cachet.

I told Ankie that it was imperative to inform my party of such a slanderous lie. Character assassination of this sort, if not stopped in its tracks, could have serious ramifications. Who would trust us if it was thought that someone notable in the Resistance (after leaving Chile, I had been made a member of the MAPU Central Committee) was working for the enemy? True, I'd promised Ankie her words would remain confidential, but this was, I said, the equivalent of someone being accused of serving the Nazis during the occupation of Holland. She had to think of it in those drastic terms, and there was also my safety and that of my family to consider, other lives that could be at risk.

Ashen, her eyes still unable to look into mine, Ankie nodded. What about the other allegations? she asked. Had my party been behind these cultural activities all this time?

Here my self-vindication began to founder. Because it was partly true that my party was the prime mover of these initiatives. My comrades provided the infrastructure in each country for the CCC, smuggled money into Chile, and distributed it, gaining power and prestige by doing so. And I wouldn't be surprised if my party had taken a small cut once in a while from the funds it transferred — as overhead, like a commission paid to a go-between.

"And you didn't tell me and the rest of the board of the CCC of this? Because you didn't trust us? Because we might not have agreed to this project if you'd told us it was surreptitiously controlled by your party, that we were being used by one group against another one, that we've been pawns in a partisan struggle? Is that why you kept me in the dark?"

I didn't really know how to answer those harsh questions. Because she was right. If I had told her at the outset, when we had first met in Paris, explained then and there the fratricidal intricacies of clandestinity, she would probably have balked, might not have handed me the funds for Chilean writers that International PEN, through Heinrich Böll, had destined for the cause.

It was true: I couldn't trust her the way she had trusted me. I hadn't trusted Max Arian, the man who had welcomed me and my family into his life and his Holland, the way he had trusted me. Impossible to trust anybody in the solidarity movement the way they trusted me. The secrecy and manipulation Ankie condemned was the way organizations that I belonged to automatically operated. In a struggle so dire and ferocious — well, dirty, ugly things were going to be done that she didn't like, that I didn't like. That's what I should have said during our first meeting years earlier, that's what I'd say in the hours ahead, the days and months ahead. At that moment, faced with her anger, her sorrow, her bafflement, all I could do was beg her to believe that I'd done everything for the best reasons, that what mattered was to focus on the wonders we'd accomplished, on the artists in Chile who had received assistance, the murals painted all over Europe.

"And what is the difference," Ankie asked, "between the way you use me and my friends here in Holland, and the use that Vrije Muziek is making of the Chilean artists? What makes you different from this Moreno man?"

I don't remember what I responded. By now I was on the verge of tears, Ankie was crying, Angélica was holding Ankie's hand. That hand of my Ankie. The first time I had seen it had been at the Place de la Bastille when she had come with her husband, Marius, to check me out, to see if Böll was right, that I was worthy of confidence. But she had already decided to love us, because that hand had been holding a bursting bunch of tulips — for your wife, she said as she leaned forward over the table of the café in Paris where we huddled after our initial hugs; for your wife, Ankie said, she must not get many flowers in these hard times. And other flowers had been left by our bed — it was her bed, she had given us her flat and her bed when we first arrived in Amsterdam, the room of her own that she needed to write her poems had been given to us for months so we would not suffer on our arrival in Holland.

That was the woman, my Ankie, our Ankie, who now finally stood up.

"Ariel, I have to present this to the board of the CCC and to other collaborators working for this . . . front your party has created. I need to be open so those who feel manipulated can decide whether they wish to continue or not."

"If you do that, it may destroy everything we've built."

"You should have thought of that when you kept the facts from us."

A scandal ensued among the Chileans, obscure and tortuous meetings, confrontations between parties in Amsterdam, Berlin, Paris, Santiago, accusations and counteraccusations, Communist friends who sided with me, but dared not do so publicly, and my party's ineffectual response to the crisis: "We're angry, Ariel, but what do you expect, for us to break up the alliance because of this? Think of the repercussions, think of the cause, man!" And of course I had to agree, the cause was more important than anything, my disarray was nothing compared to what was happening in Chile.

More devastating, however, was the split that arose among the Dutch. One of the reasons I had kept them unaware of all the political tribulations was in order to insulate them from the bickering and feuds among the revolutionary groups — and now they were themselves in danger of being torn asunder. Those who felt betrayed, like Max and Ankie, were pitted against another faction, led by my stellar students, Eric, Dick, and Emmy, who put their mentor's furtive behavior in perspective, measuring my acts against the meaningful contributions that had been made to the struggle. Bert Janssens, for instance, another of my protégés, who had organized a trip to Chile with the prestigious brass band De Volharding, to celebrate International Workers Day on May 1 in Santiago, and was arrested and deported — for him, it was obvious that our work had been wildly successful. On the other hand, Arie Sneeuw (my pal and translator, whose wife had finagled our apartment, complete with cats, for us) withdrew from solidarity work altogether. We hardly exchanged a word after that — this from the professor who had pressed the University of Amsterdam to create the post for me!

After several stormy and painful sessions, the CCC decided to ask me to remove myself from the board and stay on merely as a consultant. The remaining members would branch out to include Chileans of all political persuasions and be extra-vigilant in how they distributed their funds. They were ready to persevere because the tasks ahead were indeed too vital to be ruined by this mess. Meetings with Ankie and Max, however, became businesslike and perfunctory, full of suspicion and regret. It was unbearable to see Max's pain. It hurt every time he would call and ask for Angélica, to find out how she was doing, how Joaquín was coming along, to set up a time for Rodrigo to play with his two sons, not asking how I was, it hurt that Maartje was warm to me each time we met but that we no longer visited her home, it hurt that she did not know how to manage this estrangement. I grieved as I saw the old friendship vanishing, and nothing I could do or say seemed to be enough to thaw the civil iciness.

I grieved for what I had lost. Holland, the haven where I had started to defeat distance again, was now a place embittered by a new loneliness, isolated from the two individuals I had thought of as my new sister, my new brother, and who now seemed to bemoan having taken me in. I felt this even as Ankie, oh so slowly, began to forgive me, started coming by for her morning tea or afternoon drink. We gradually became the mischievous schemers of yesterday, allies in the struggle for poetry, as we explored together the traps of translation. She smiled again, she sat back in the chair, she accepted some sherry, I told myself we would soon again be like two siblings, even if once in a while her eyes would haze and blur over with a quick pang of wariness. I was discovering that trust, once lost, may take a lifetime to rebuild.

As for Max, his misgivings were less easily assuaged. Eventually, though, thanks in large part to the stubborn women of our respective families and the immense goodness of his heart, Max allowed me to start healing our wounds, so by the time we left Holland, in the summer of 1980, the rift had begun to mend. And five years later, when we returned for a visit, Max and Ankie were among the two dozen friends waiting for us at the airport at three in the morning with flowers and chocolates, verses and bear hugs. As I write these words in 2010, I am preparing to visit Max in Amsterdam and to spend every hour I can walking the streets with him and celebrate the marvels of his seven grandchildren. It warms my heart to remember how he and Maartje journeyed just recently to London to attend a workshop production of my newest opera, *Naciketa,* and I can safely state that we have become even closer than before. And Ankie, we will do our best to visit soon the town in the south of France and its cascading river where her ashes are scattered. I feel her as deeply a part of my life now that she is dead as when we shared so much life and laughter and solidarity.

So we managed, Ankie and Max and I, with the help of Angélica, to mend the fracture in our lives, but not before the loss of those two soul mates, temporary as it might have been, shook me to the core, forced me to face misgivings about myself that I had been postponing for far too long.

I had left Chile with my hands empty, riddled with guilt and the need to send back assistance, I had come to Holland as a beggar, I had wandered abroad and told myself that I had the right to employ every weapon at my disposal, all my talent, all my magnetism, all my engaging enthusiasm, the tools of my trade, words, words, words, I told myself that it didn't matter, a little lie here, a small falsehood there. It didn't matter because it was all for a good cause, it didn't matter because I was a victim, and victims are always right, it didn't matter because those people who were receiving us had a home and we didn't, they had security and we didn't, they had democracy and we Chileans didn't, they had prosperity and we didn't, they had contacts and we didn't.

The pattern had started in France and persisted in Holland. I had fled from Chile, my life filled with uncertainty, except the certainty that I was on the right side of history, shielded by the mere existence of a fiend like Pinochet from reservations about my own ethics that had buffeted me since adolescence. When you are struggling against a dictator who betrayed the president who named him to his post, who bombed the presidential palace and abolished Parliament, it is not hard to feel good about yourself, to suppress any qualms. Things go wrong? Just blame the dictator. It became a sort of joke — *hijo de puta Pinochet,* I kept cursing him for what he did and also for what he could not possibly be responsible for. I can remember an evening in Holland — it was before Joaquín's birth, before all hell broke loose — when our old car conked out in the midst of a blizzard. It was my fault, I hadn't changed the oil in months because I hated having a car, even if it was a gift, hated looking even a bit privileged, and I turned to Angélica and snarled *hijo de puta Pinochet,* if it hadn't been for him

I'd have remembered the oil for the engine, and we both laughed. We laughed at how absurd we were. Thanks, Pinochet, thanks for offering us an excuse not to look at our own shortcomings, thanks for safeguarding us from having to face them, *gracias, General.*

My laughter did not last long.

Here was the truth about myself that I could no longer escape: I had misled my friends, I had exploited their love, I had used their confidence in me, and I had done so willingly, almost daring retribution. Somebody inside my mind had calculated every move, swept aside the Ariel who loathed this masquerade, I had looked out on the barren, welcoming world and did not see the others for what they were but for what they could give to the Resistance, what I could extract, I measured their utility in a game I felt I controlled, the only thing I seemed to control, I saw them as objects, things, hostages of my charisma. Not only after the coup, when the going was rough, when everybody else around me was doing it, all of us survivors clawing our way out of the slime in any way we could, hogging any help available, treating those miracles of solidarity as if they were merchandise, names on a list — that attitude had started way before the coup, that cold and scheming voice had been my companion for as long as I could remember, perhaps since I had outlasted the bout of pneumonia in that hospital in New York. That scared child was still there, battered and abandoned and ready to do anything to keep breathing, a malevolent voice I shunted aside, struggled against, tried to suffocate with my shame. But it was always there, someone always there enjoying that power, kept in abeyance when things were going well but ready to come out and play, those evil spirits, after the coup, because then I had found myself suddenly vulnerable again. My remorse demanded that I prove that I deserved to have lived to tell the tale, more in need than ever of that jolt of supremacy, that sway of entitlement, more in need of the spotlight of success, bolstering my crushed ego with the trophies of my conquests, a Don

Juan of the solidarity world, desperate to further the revolution and be rid of Pinochet, and not caring, not caring enough, whom I damaged along the way.

The year 1979, which had opened so auspiciously, with birth and rebirth and the fall of Somoza in Nicaragua and the Shah in Iran and dictators ousted everywhere, had gone dreadfully wrong. That terrifying year was a turning point: from that time on I would be unable to pretend that I could cross through the fire of exile and violence without being scorched by it. Nobody can survive pain and defeat and cruelty on that scale, nobody can lose their home and their land and their friends, and remain pure and immaculate.

So when I emerged from that calamity of self-scrutiny in Amsterdam, something had, I think, radically changed inside me. I am not going to claim today, I will never again declare or pretend, that I am incapable of manipulating people or deceiving them; I can't presume that I will never betray those who have put their trust in me. But that's the lesson I learned, that I hope I learned, from the trials I went through: I am not a hero, not superior, not much better, in fact, than anyone else on this planet. The monster who has been my close companion longer than I can remember has not vanished; it is still there, still would like to come out, still whispers a slither of words to me from time to time. I can only hope, can only pray, that to acknowledge the darkness is already a way of vanquishing its dominion, that to accept your mortality is already a way of preparing for a good death.

It was out of that darkness and imperfection that *Death and the Maiden* emerged, my urge to demand that the audience, demand that my country, go through its own inferno, ask itself the basic question of what we were to do now that Pinochet had retired from the scene, but not his shadow, not his shadow, what were we to do now that he was no longer there to guide us with his demonic, grandfatherly eyes, his mercenary heart.

Fragment from the Diary of My Return to Chile in 1990

AUGUST 28

Today I received a letter from Eric Gerzon.

We had not known of his existence, had never set foot in Amsterdam, that failed saint's day party of Angélica's in September of 1973 when death had come calling for so many of our brothers and sisters, and yet he is now as much a brother, perhaps even more so, than the friends I have here in Chile, that's what I need to admit. Because death had also stalked us abroad, cast us into a chasm of solitude and grief, shown me the worst face in the mirror, shattered everything I believed in, brought me to the brink of drowning, and in that dark night of the soul, Eric had been there, caring for us with a love as unconditional as a mother's, with the sort of infinite eyes and soft hands and benevolence that we needed more than any daily bread.

And now he is in Holland and he writes to tell me he is planning a trip to visit us in Santiago soon and will stay with us here at Zapiola and in the midst of my jubilation, my many hallelujahs that brought Angélica running to hear the good news, something strange twists inside me, a reversal of time, a dislocation of space. I am briefly returned to those moments abroad when I would receive letters from Chile and examine the flimsy envelope in disbelief, turn it over and over, try to extract all the life for which those insufficient words must compensate. As I read Eric's message from Amsterdam, I remember the hours in exile deciphering between the lines, excavating for implications that may not be there, construing what is inexorably left out, the glory of gossip, the flow we are missing.

And so it starts all over again, the distance. I am in Chile and Eric is way over there, on the other side of the world, eight days separate me from this man whom I love as much as I love Queno or Cacho or Susana or Pepe, this gentle soul who has helped me to survive and come back to this land where I miss him, where I will soon greet him in our house as he received us into his heart. Max has also announced a possible visit, and maybe they will all come, Ankie and Deena and Maartje, Pat and Mark and John, and Jackson has told us that he's coming for the Amnesty concert in October. I hope they all come, from Paris and Los Angeles, from Amsterdam and Washington and New York, one by one and all together, so we can give back to them the love they gave so openly, never expecting anything but our joy and our return to Chile.

Jubilation, yes, hallelujah, I await that dawn of their visit.

But there will never again be a day in my life when all my friends from Chile and from abroad are together, when the diaspora and the scattering of seeds will end, there will never again be a day or a night, there will never again be a simple gathering that is not inescapably and forever incomplete.

The doubleness of exile has followed me home.

•

SPEAKING of doubleness.

A rebellion is brewing.

My two languages have stood by patiently as I traced a journey from Buenos Aires to Paris to Amsterdam, and now that this memoir has reached the end of our European odyssey, now that I am about to disembark, quite literally, on the shores of the United States, both languages are demanding attention, that I bring them into this narrative. Remember, Ariel, *recuerda*, this is also the story of how someone who left Chile *proclamándose* a fanatically monolingual *escritor* became instead a duplicitous adulterer of languages, in love with two equally exacting tongues.

Exile did it.

The language of Keats and Lincoln had been awaiting its

chance ever since the day in 1969 when I abjured any and all allegiance to it, that day when I renounced the vernacular in which I had built up until then the house of my identity, because I felt English too tainted by empire and privilege and distance to convey my colloquial Chilean experience of love, literature, and revolution.

Maybe English's forbearance derived from the knowledge that the odds were in its favor, that in this increasingly globalized world, and given my increasingly unhinged life, I would finally fall back into its arms.

Not that I was aware of the strategy that the jealous and mischievous jargon of Mark Twain and Groucho Marx and Monty Python was hatching when I left Chile. As soon as my exile commenced in Buenos Aires in 1973, I used English to speak to foreign correspondents, connect with solidarity people in Europe, and obtain the first funds for cultural resistance, giving the matter not a first thought, not a second one, less consideration than that due to a toothbrush, a mere instrument to get the job done.

In France, the syllables of Shakespeare did not tempt me, and were only employed for persuasive purposes on trips away from Paris, in Berlin, Copenhagen, Rome, The Hague, certainly in London and Oxford. I would never have dared to brazenly venture into the corridors of power if I had not been accompanied by that lingua franca. I can recall a trip to Stockholm in the spring of 1975, hoping to induce Olof Palme, then Sweden's prime minister, to lend us a boat on which we, the artists of Chile, could set sail for Valparaíso, daring the junta to stop the landing. A harebrained scheme, but feasible enough to have enlisted many Chilean cultural figures, as well as García Márquez, generally a cooler head than mine, who had called Pierre Schori, Palme's right-hand man, to set up a meeting. The prime minister listened to my proposal, wished us well in our struggle, adding that he'd get back to me.

The call never came. Not that I was surprised. As Schori (who would become a friend years later) escorted me out, he said:

"That's the most irresponsible political plan I've ever encountered. We could be sending your greatest artists, the most visible face of your country abroad, to their doom." I should have bowed to his superior wisdom, but my eloquent English was there to articulate all sorts of reasons why my crazy scheme made sense.

No such expressiveness available to me back in Paris.

I had loved French ever since I had heard its precise sonorities as a child of nine, meandering along Parisian *rues* holding hands with my dad and mom. A decade later in Chile, Balzac and Proust and Simenon introduced my older adolescent self to *quartiers* and *bistros*. Baudelaire had made me aware of the smell of those *boulevards,* and Zola had described the vegetables and meat and female flesh, the arcades and window displays. And the paintings of Parisian sunsets, the wispy chalk-colored clouds, the Seine, the cathedral, the Gare d'Austerlitz, the Rue Mouffetard. Truffaut and Godard and Jean Renoir and *Les Enfants du Paradis.* And history: I knew more about the French Commune than about the independence of Chile, more about Marx's *Eighteenth Brumaire* than about the history of Ecuador.

Yet that very familiarity, the moveable feast that Hemingway had immortalized as the lot of the artist in Paris, and to which I was not invited, made my life all the more despondent and ghostly, blocked any chance that I would welcome French or that French would welcome me. I had not come to that city to worship at the shrine of Beauvoir and Rodin, was too full of darkness to be edified or illuminated, too aggrieved to lap up eternal sagacity, Camus-like, at Les Deux Magots. Not that I didn't give it a try.

A bit after my arrival in Paris, Julio Cortázar introduced me to Michel Foucault, one of my academic idols, and I had been embarrassingly tongue-tied, unable to vent the ideas about repression in Chile and its connection to his Panopticon coursing through my overwhelmed brain. And then, a month or so later, Jean-Pierre Faye, a French intellectual sympathetic to the cause of Chile, offered to set up a meeting with Sartre.

During my late adolescence in Chile and all through the subsequent years of my young adulthood, Jean-Paul Sartre had been a guiding light. More than anyone else, he had popularized the existentialism that was all the rage back then, giving it an ethical twist that appealed to many of my generation worldwide, especially during the sixties, when Sartre opposed colonialism and the Vietnam War. His conviction that avant-garde writers could not be indifferent to the fate of the world echoed my own commitment to drastic change in Latin America, as did his proclamation that every aesthetic choice should entail a political act. By combining essays and journalism with pathbreaking plays and novels, he offered himself as a model, the intellectual *engagé* whose partisan struggles could not be separated from art.

Despite all this, I decided not to meet him when the chance arose. I couldn't stand the idea of stammering my broken, clumsy French to the man who had contributed so brilliantly to my own ability to analyze the world with sophistication and elegance, who had provided me with the somber vocabulary with which I had defined liberty, alienation, being, nothingness. I wasn't going to risk duplicating with Sartre the awkwardness that had beset me when attempting to communicate with Foucault. I had been carrying on a dialogue with the great Jean-Paul for years, addressing him in my mind with the most exquisite grammar—and I preferred to keep it that way. Someday soon, I lied to myself, my French will be adequate enough for that momentous encounter.

But my French only got rustier with time. I would be in a solidarity meeting—discussions with comrades, with French men and women more than willing to overlook any morphological gaffes—and find myself stopping in the middle of a pidgin phrase, flailing for the *mot juste*, for any word at all, *juste* or *injuste*, searching for the right nuance or Left Bank allusion, until finally, after a discomfiting interlude, I would slip into a mix of Spanish and pseudo-French, with a dash of useless English thrown in for good measure.

What made that incapacity to express myself especially un-nerving was that it had never happened before, not to an Ariel so fluent in two tongues. My bewilderment had begun the day I alighted in Paris and lost my way, and the man I approached for directions repeated my words back to me, correcting my accent as if he had been deputized by the Académie Française to avenge any deviance. It was the first of many such icy stares and enunciating lips, and as similar incidents multiplied, I felt, well, terrorized. A word I use here with caution, given the real terror that was ram-paging through Chile and Argentina and Guatemala. Yet the help-lessness I felt was real, the distress all the worse because it could not be communicated.

In Chile we have a saying, *En boca cerrada no entran moscas*, Flies don't enter a closed mouth. I'd take that advice, avoid any chance of being scorned in the blasted heath of exile. I would often ask Rodrigo — whose French eventually became so proficient that it would have made Corneille rise from his tomb and applaud — to go shopping with me. One deaf-mute day I was so overcome with bashfulness in a butcher shop that even my Spanish abandoned me. Could this really be me, stuttering as if my lips had a conta-gious disease, my finger pointing at a slab of meat, dependent on my son's flawless diction?

Proud as I was of Rodrigo's linguistic accomplishments, grate-ful that he frequently salvaged me from mortification, I also stifled him, as virulently as I had my own tongue. When he became ever so slightly boisterous or raised his voice excitedly in a public place, I would hiss at him to keep quiet. I wanted no trouble. I couldn't stand the idea of facing a group of neighbors or French women in a park or a playground, couldn't bear my impotence, not be-ing able to defend our boy with the subtlety and irony I no longer commanded. Don't call attention to yourself, Ariel. Behave, Ro-drigo. They can do anything to us, my child, these people in their own city, anything they want.

So this is how countless dispossessed people must feel each

morning, deprived of the instruments that might buffer them from authority; this is what it means to be a stranger in a world that does not belong to you, powerless in Gaza, blind in the City of Light.

No wonder Holland was such a relief.

It didn't matter that Dutch was an absolutely extraneous tongue or that I did not master it during my four years in the Netherlands. It welcomed me like Amsterdam itself.

Proof came by November of 1976, our first year there. As the days grew shorter, I noticed a strange frenzy on the streets of the city: the Dutch spellbound by a writing fever, as boys and girls, mothers and grandparents, scribbled on notepads, the city brimming with frenetic bards on multicolored trams and violet buses and in cafés. I remember an old woman straight out of Rembrandt sitting at a large bow window surrounded by flowers, jotting something down, muttering to herself, chuckling.

"It's for Sinterklaasje," Max Arian elucidated, inviting us to accompany him and his family to the Leidseplein on a Saturday in mid-November, where the overflowing, apple-cheeked crowd cheered the arrival of Saint Nick as he descended on a great white horse from the barge that had brought him all the way from Spain. While this Dutch version of Santa Claus reverentially received the keys to the city, his helper, Zwarte Piet, in blackface and frolicking medieval-jester clothes, threw candies to the children. This was merely the first phase of a celebration, Max explained, to be continued with us, he hoped, on December 5, the saint's birthday, when gifts would be exchanged. Offerings should be playful — hiding the tiniest contribution in a colossal box or smuggling a trinket into somebody's pocket or even into, say, a potato. Anything would do, provided it was accompanied by a bit of doggerel written in Dutch, the funnier the better.

Back home, practical Angélica decided she'd shop for the presents while Rodrigo and Ariel would supply the limericks. Over the next few weeks, I joined the ranks of the lyric-mad Dutch and churned out a variety of couplets, laughing my head off with Rodrigo as we misused and misrhymed Dutch words, paired them with outlandish Spanish and English endings, a blissful, carefree grammatical mess, a respite from the poems of the *Desaparecidos* soberly accumulating on the all-purpose green table bought at the flea market. Our lyrics were not something I could have shown to Sartre, discussed with Foucault, or presented to Olof Palme, but what a joy not to be judged, not to worry about being lucid when we came upon the word *lekker,* meaning good to eat, scrumptious, and immediately rhymed it with Woody Woodpecker, to accompany a bag of sweets for the kids.

And that was as far into Dutch as I needed to venture.

I connected with most people in Holland through my English, avoiding the syntactic gymnastics that had disconcerted my French interlocutors. That's all right, I said, as English again became part of my everyday experience, almost a second skin. Spanish, the language of my writing, was still dominant at the university and in the solidarity movement and, of course, at home.

Nevertheless, I would lift my eyes from my stories or poems or the *Widows* novel I had begun, and Rodrigo's voice down in the street, playing an improvised mini-soccer game with his Amsterdam friends, would yank me back to the here and now where the invented Chile began to vanish, leaving behind it a world of frontiers and oceans and border guards. What hit me at those moments like a soft storm was the real geographical distance Pinochet could punish us with, the fact that Rodrigo was in danger of losing not only the country but the language that was that country's ultimate guardian — that what had happened to me as an infant could happen to him as a child.

Languages are built on shared silences, assumptions never spelled in dictionaries, what we omit, fail to explain, because we're often unaware that an explanation might be required to clarify what we mean. One day, Dorothée, a student at the University of Amsterdam who had been translating an article of mine about Chile's Disappeared for a local paper, came with a question. "There," she said, jabbing her finger at a paragraph. "*Hay una contradicción.*"

I could find nothing wrong with the offending phrase, no contradiction. It claimed that dictators want to sweep people from the minds of humanity, store them in an archive in order to forget them. "That's the word that doesn't work," Dorothée insisted, pointing to the Spanish word *archivar,* meaning to classify a document in an archive. For her, when you officially put something away, you're consigning it to memory, making it retrievable. If the State, *el Estado,* wanted to obliterate opponents, as in Chile with the *Desaparecidos,* she said, then it would obviously take them *out* of the archives. As a Dutch citizen, she expected public servants to preserve an agreed-upon past, which existed as irrefutably as the dams that kept the sea at bay. Whereas for most Latin Americans anything filed in a public archive is secreted by an adversarial and shadowy State that you should never trust, anything filed away is on the incessant verge of oblivion. As for the past that Dorothée considered static, I saw it as a chain of shifting events that nobody had really registered, that each generation had to reinvent anew.

So how would Rodrigo look at *el Estado,* interpret the Past? Like the Dutch children in the street or like his Chilean parents watching him from the faraway windows? How would the kids born in Valparaíso but brought up in Maastricht name the night, now that they were under a northern sky with no *Cruz del Sur* to guide them, now that the land of their forefathers was only a faded poster on our wall, remembered a bit less each evening, mispronounced a bit more each morning?

Suddenly, from the most disparate and scattered countries, a number of parallel initiatives were begun: community centers that had been dedicated to aiding the Resistance started to concentrate on the refugees themselves, offering classes on Chilean history and literature, guitar and painting lessons, soccer tourneys and summer camps, anything to keep the kids in contact with each other, reestablish links to a forgettable land. The problem was that Rodrigo was too tired to attend the evening or weekend sessions held in Rotterdam, as five days a week he took a bus to the only French school available, a two-hour bus ride to The Hague each way. A tremendous sacrifice for him, and a considerable cost to my parents, who were footing the bill, but this at least guaranteed some linguistic continuity in his studies. We figured there'd be some *lycée* in whatever country we ended up next.

The diabolical plan that I hatched to fortify his ties with Spanish was for us to read together for an hour before going to bed. One page by the son, one by the father, and so on, until we finished the chapter. The books selected were not political or crammed with Chilean history. Rodrigo and I spelled out each evening the breathtaking wanderings — not like ours, full of supercilious cats and an occasional colorful Sinterklaasje — of the great Sandokán, the magnificent *Tigre de la Malasia,* defender of cringing damsels in distress and people enslaved by sadistic despots. Those adventures had not even been written originally in Spanish, but scribbled by the Italian pen of Emilio Salgari, who had thrilled generations of readers in the late nineteenth century.

Salgari's tales were a strange choice for an intellectual who had gained notoriety by attacking superheroes who swung in from the outside to resolve dilemmas for the passive masses, a writer whose most famous book up until then was *How to Read Donald Duck,* with its denunciation of popular-culture imperialism. After all, Sandokán disputed my theory of history as the manifold creation of everyday human beings. But I was moving — as were many on the left — towards a certain pragmatism. What mattered was to keep Spanish from being quelled in the archives of forgetfulness. Because not only people but syllables and significance too can be abducted night after night and never heard of again. And so Sandokán wrestled with thugs and pirates, avoided being trampled by runaway elephants or consumed by a swamp of alligators, while Rodrigo labored with his tongue and his teeth and his saliva for the right to hold on to his faraway friends Cervantes and Bolívar; be able one day to read the prophecy of Salvador Allende that the *alamedas* of history would open and through them would walk the free men of tomorrow; prepare himself for Quevedo's *Miré los muros de la patria mía* and García Márquez's *Muchos años después, ante el pelotón de fusilamiento,* the walls of the forbidden fatherland and the years gone by and the firing squads and *verde que te quiero verde* of Lorca, I wanted him to dream the color green in Spanish, I wanted to dream that color myself only in Spanish. I grew more intent on this the closer we got to our impending trip to the United States — even if it was only for a year, we told ourselves, if only on the way to a presumably more permanent home in a Spanish-speaking country like Mexico. Anxious, nevertheless, because I was finally going to live among people who spoke a language that, unlike Dutch or French, I could not easily deny.

The language of the enemy, of the men who had destroyed Allende.

I had visited the States twice during my European exile in order to help organize cultural solidarity with Chile. Strange to be back in the land that had nurtured my childhood, had given me Lou Gehrig and the Bobbsey Twins and Eartha Kitt's "Santa Baby" and *Miracle on 34th Street* and Glenn Ford on the 3:10 to Yuma and *The Fire Next Time* and the Mars bars this time. The country of the Underground Railroad and Martin Luther King Jr. and César Chávez and William Sloane Coffin and Teddy Kennedy, Tom Hayden and Studs Terkel and Susan B. Anthony. But also the country of Nixon and Kissinger and the marines and the School of the Americas. I felt nauseated walking past the façades of buildings in downtown D.C. where conspiracies against President Allende had been forged, but I was assuaged almost immediately by Ella Fitzgerald crooning a memorable Gershwin tune on a street vendor's radio. Something always came along to remind me of what I loved about America, Emerson asking a jailed Thoreau what was he doing *in there* and Thoreau's response, no, the question is, what are you doing *out there,* and the Carter administration confronting the Argentine military. But then my gringo heart would turn to ice when I read the Church committee's Senate report on the destabilization activities of the CIA, the two Americas duking it out for my soul. And as I articulated our cause to congressmen and authors, lawyers and film stars, I also realized on those two preliminary visits how well I fit into this America, how deeply I understood its mind and its motives, the instantly accessible references and metaphors — what other Chilean could quote Will Rogers and Wilt Chamberlain, Emily Dickinson and Charlie Brown? Good grief, I saw a hint of my future as a bridge between continents and cultures, a glimmer, a glance, an insinuation in the landscape of my spirit, and then . . .

No, not the States. Even when I accepted a one-year stint at the Smithsonian's Wilson Center to write my next novel, *La Última Canción de Manuel Sendero,* I did so only because we would hurry on to Mexico once that year was done. The sojourn in D.C. was possible psychologically because I refused to entertain a more permanent residency in Norte América. I think I was afraid that, once installed in the country of my childhood body, I would stay there forever. Or perhaps simply afraid of having to coexist in such promiscuous proximity with neighbors who could be friends, could be enemies, might conceivably be both friends and enemies.

Vamos a hacer el cruce por mar.

That's what I had told Angélica one day as we contemplated the upcoming 1980 trip to Washington, that we should try to make the crossing by sea, an arrangement consented to by the Wilson Center, a way of solving a problem that had crept up on us almost unawares.

Books. Four years with a salary had stubbornly replenished a library that needed, along with some personal effects, to leave Europe with us, and a sea voyage could provide that transportation free of charge. And such a drawn-out passage would give us time to adjust, to extend a barrier between the seven European years and the uncertain future.

It was only after the cargo ship set sail and we had bid farewell to our Dutch friends on the dock, only when I saw the coastline gradually receding behind us, that I realized there was an involuntary meaning to our choice. Next to our Amsterdam friends waving goodbye were my mother and father, who had come to help us make the move. Those very parents had, each separately, taken this Atlantic journey from Europe so many decades ago, this was the route of their families when they abandoned the Old World at the turn of the century. This phase of my odyssey could be seen as renewing a cycle my forebears had started seventy-some years back, that Angélica's ancestors had experienced in the eighteenth and nineteenth centuries.

My mother had been three months old when her parents had set sail from Hamburg. "Worse than third class, we went," my grandma Baba Clara once muttered to me in her broken Spanish and then lapsed into the Yiddish I did not understand. I could picture the stinking hold, the heaving waves, the men segregated from the women, the smell of unwashed bodies, and the mix of languages, to be stoically endured as long as they were far away from the next pogrom. My grandmother's father on my mother's side, a cattle dealer, had been killed by the Cossacks in 1905 and his house looted. The rest of my mother's Weissman family had found safety in a nearby church, most of them escaping from Kishinev a few years later. Across the Black Sea in Odessa, where the Dorfmans lived, the murderous hordes had been resisted, mainly by contingents of Jewish brigands, thieves, and smugglers, the sort that populate the stories of Isaac Babel. So my father's family had not migrated to Argentina in order to flee massacres, but rather in search of economic opportunity.

My new ritual of expatriation decades later might not have pleased my grandparents. They had emigrated so their offspring would never have to negotiate oceans ever again—and here I was on a sea that belonged to no one, crossing between countries neither of which was my own, crossing with the memory and nearby ghosts of my tribe. But they might have understood the pioneer excitement of starting again that I felt, the thrill of beginning from scratch.

Well, not quite. There were our boxes of books. Deposited on a wharf in Baltimore and no way to get them to Washington, D.C.

It was already a miracle that we had been able to sneak our belongings into the United States at all.

"Forty-seven boxes?" the customs agent asked. "You're here on a temporary visa and you're bringing in forty-seven boxes and sixteen pieces of luggage?"

I explained that this was just a stopover on the way to Mexico.

"Mexico? Have you got a visa for Mexico, a job offer?"

"Several offers, but I haven't decided which to accept. We'll travel there next week—here are our plane tickets—and then I'll make a decision."

"And what's in the boxes?"

"There's a sewing machine and some toys and—but mostly books. I'm a writer."

"You're a writer?"

I looked at him. He was tall and had a monumental mustache. There was a hint of mischief in his eyes and the intimation of a smile at the creased corners of his lips. I took a leap of faith, let a litany pour out: "A writer, yes. One of the things I want to do once we're settled in is come back to Baltimore, you know, to visit Poe's tomb. I've always loved his work, a shame that he died so young, but if he had to die, this was the right place, halfway between the North and the South, because . . ." and I just couldn't stop, out of nervousness, my motor mouth. "Poe was torn between the cold detective intelligence of the future and the gothic horrors of the past. In fact, in those boxes there are two volumes of Poe's short stories in Spanish, translated by Julio Cortázar, the Argentine writer—"

"*Hopscotch,*" the customs agent said unexpectedly. "Yeah, I read that one. A bit difficult, but I liked it."

"He's a friend," I said, too quickly. "He even wrote a dedication to us. If you want, I can show you the—"

"Have you read *The Flounder*?" the man asked. "By Günter Grass. That's one of my favorites."

Angélica, who tends to be quiet, suddenly piped up. "We know Günter Grass. We visited him at his cottage outside Hamburg," she said, without telling the whole story of that visit.

"Is that so? They say he's one helluva cook."

"He cooked for us," I said, not telling the entire truth but not exactly lying either.

And that was that. The customs officer grabbed our declaration, stamped it, and gave us instructions on how to find Poe's grave—it was hidden away, he said, but well worth a visit.

Literature had got us into Baltimore, but literature could not get us from Baltimore to Washington.

For that, we needed a credit card.

I had told Angélica that renting a van would be a matter of calling up one of the many companies and we'd be on our way, and for some reason my wife believed me. Once in a while, I guess, she gets tired of contesting the breezy confidence with which I affirm or prognosticate something, and gives me a pass. She should have been more dubious.

Not one rental company would let me lease a van. Not Avis, not Hertz, not Budget or National. No credit card? No U.S. driver's license? I offered to leave a deposit, I had cash, I—my pleas went unheeded. In America, if you don't have a credit card, one of the agents bluntly told me on the phone, you don't exist.

I was in a small shack at the end of the dock, where an amiable official shrugged his shoulders in sympathy each time I hung up.

Outside I could see Angélica sitting on one of the boxes, and Rodrigo running around with Joaquín on a leash, harnessed up so he wouldn't jump into oil-streaked Baltimore harbor. The sky had blackened, and soon it would be raining on us, on our boxes. Not raining. Pouring.

"I have to change Joaquín's diaper," Angélica said.

I tried calling my friend Pat Breslin, at whose home in D.C. we were going to leave our stuff until our return from Mexico. No answer. I called Art and Joan Domike, friends of my parents from Chilean days. Nobody was home.

Things did not improve as Rodrigo and I scoured the area in search of a savior. Not far from that dock, on an election night in 1849, Poe had staggered through the streets of Baltimore, taken from bar to polling place to bar to polling place so he could vote over and over again, and I felt like invoking the purported last words of Poe, *Lord help my soul,* but it was not the Lord who came to help us, though who can tell why and how miracles happen? In this case, the miracle came in the form and shape and name of a hefty man called Witherspoon. He was a stevedore just finishing his shift, and sure, he had a van, and sure, he'd take us to D.C., his brother could help with the boxes, those over there, yeah, they fit in my van, you all gonna fit in my van, and it would be, how about a hundred bucks? His smile was as broad as the horizon and far more appreciated, because by then we couldn't even *see* the horizon, that's how dark it was, and the miracles kept on piling up.

Just as the last box was muscularly shoved inside the van, the tempest burst and we were snug and safe, Witherspoon at the wheel and me next to him with Joaquín on my lap and in the back Angélica and Rodrigo and the sanctified brother of Witherspoon, and we were off and it didn't matter that we got lost on the way to D.C. because our driver told us the story of his family, their own struggle against homelessness. We had been saved, it turned out, by the descendants of slaves who had come here in the hold of a ship, and not in any stateroom with an ocean view.

Next we were greeted by the son of Irish migrants, Pat Breslin, and his delightfully pregnant wife, Janet, neither of whom seemed to mind that we were arriving so late at night. Soon all our boxes and luggage were in the basement, and our stevedore pals from Baltimore demurred when Pat offered them a bite, our saviors had to get home before their wives began to worry, and I watched them recede into the night and wondered again at how compassion works its magic and realized, as we sipped the excellent Chilean Concha y Toro *tinto* that Pat had just uncorked, that we might not have a credit card but, man, did we exist, and it was only after all the commotion subsided and I was able to relax, only then did it come to me that I had not yet had the time to absorb or even register that I was back, I had returned to the country of my childhood, and that this welcome augured well, that it had seemed as if the demons of exile were going to play yet one more drastic trick on us, and instead, as always, even here in the States, the angels of solidarity had won the battle.

WE'D NEED them in the United States, those angels, that solidarity, sooner than we thought.

Not that I expected trouble.

I had left nothing to chance regarding our next destination, the *México lindo* where we intended to wait out Pinochet and bring up our sons in Spanish. I'd nailed down every detail on two trips down there. The first had been with the whole family in August 1980, when I'd been invited as a juror of a literary prize, jointly sponsored by Nueva Imagen, the publisher that was bringing out my new books of short stories and poems, and *Revista Proceso*, a Mexican magazine headed by the inimitable Julio Scherer. On the last night of our week together in the resort of Cocoyoc, under the shadow of the volcano that had goaded Malcolm Lowry into madness, I popped a question to Scherer at the instigation of my co-jurors García Márquez and Cortázar, both of them sure he'd relish helping us avoid the hell that exiles in Mexico went through each year in order to renew their immigrant status, I asked Scherer if he could find a way to obtain a permanent visa for us to reside in his country as of September 1981 and thus bypass a callous bureau-cracy.

"Consider it done," he answered, adding, with his usual gener-osity, "I'll fly you down here to make the arrangements."

February 1981 found me in Mexico City for a second time. My first evening there, Julio Scherer told me, over dinner in a small restaurant off Insurgentes, that it had all been taken care of. "I saw the president yesterday," he said, "and López Portillo asked why Ariel Dorfman was not already here yesterday, the day before yes-terday, *que venga ya mismo.* The president gave instructions to *Gobernación* to expedite permanent residency."

The next day, I reached an agreement with Juan Somavía to work on a project on alternative media at the Latin American In-stitute of Transnational Studies, ILET, which he was then heading. Juan was glad to circumvent the ruthless corridors of the Mexican government in order to secure my status — if Scherer had settled the matter, there was no need to worry, an opinion seconded by other Chileans. I can remember a lunch with Tencha Bussi, Allen-de's widow, where she agreed it was unnecessary for her to inter-cede. If Julio Scherer is involved, she said, you're in safe hands, he is a man of his word, *un verdadero caballero.*

Elated by all this, I returned to Washington, where we visited the Mexican consulate with our photographs and the ILET con-tract. I made a call to Mexico in late May 1981 to an imperturbable Julio: everything was fine, our permit would arrive sometime in June, no need to worry. I had averred the same to Angélica, noth-ing to worry about, *no hay para qué preocuparse, mi amor,* my love, it's all taken care of, you know how these things take time. We had given notice on our rental in Bethesda, and the owners were al-

ready showing it to possible new tenants and Rodrigo had reluctantly bid farewell to his *copins* at the nearby Lycée Rochambeau,
his grades sent on to the French school in Mexico City to guarantee a smooth first year of transition, but still no visa, not yet, the
functionaries at the consulate were stone-faced, claimed zero information, come back *mañana* and *mañana*, always followed by
another *mañana*, as if we were trapped in some Carmen Miranda
song.

A few nervous calls to Scherer and he repeated that it was all
under control, what could possibly be amiss? So we repacked our
bags and boxes, every last item stored away by the indefatigable
Angélica and ready to go — and off we went to Canada in mid-July
of 1981, a much-needed holiday paid for by my ever-willing parents.

As I drove the rented Chevy towards Prince Edward Island,
with my dad by my side and Angélica and my mom and the two
boys in the back, I mused to myself that it was odd that I hadn't
been able to connect with Julio Scherer before leaving on our
vacation. But I didn't dare to express my concerns out loud, everything was on schedule, his secretary said that he said, and so
I needed to be patient, *no hay que preocuparse.* Just then, I can
remember seeing, out the car window, a scene from an Edward
Hopper painting, or maybe an Andrew Wyeth. There, standing by
itself against the Canadian prairie, reflecting the blue luminosity
of the lakes and reverberating sun, was a solitary telephone booth,
and something in my foot told me to start braking, and the car
obeyed me and stopped next to that phantom booth.

So eerie is that memory that I tell myself now that I must be
inventing it, maybe I actually made the call to Scherer from a gas
station, maybe it was that night at our hotel, and yet this is how I
remember it: I dial the number in Mexico City and slip coin after coin into the slot and rove my eye from the tundra-like landscape to the car, filled with the five people in the world I most love,
and then Julio is on the line, Julio Scherer is admitting, he is tak-

ing his time about it, he is hemming and hawing, he is embarrassed as hell, doesn't know what to say, but ekes it out anyway: *Estamos jodidos*. We're screwed. *Proceso* had run a series of articles about corruption in the oil ministry and the furious president declared war on the magazine, pulling all advertising from it, just as Echeverría, the previous president, had declared war on *Excelsior* when Scherer ran that newspaper, and now history is repeating itself. López Portillo will try to starve *Proceso* into submission, deny any requests in a land where all power stems from the Imperial Presidency and the governing party, the PRI. López Portillo has said that all favors afforded to Scherer have been canceled and specifically mentioned that no special treatment be accorded to Scherer's friend Dorfman.

Julio is desperate with shame and rage but there's nothing he can do. He can't stop publishing articles about the oil industry; the most he can offer is that we travel on a tourist visa and he'll contact someone he knows at *Gobernación* to see how to navigate the Kafkaesque bureaucracy, but everything will turn out all right, this confrontation can't last long, this president or the next one will relent and we'll negotiate a permanent visa, my work permit. Scherer feels that his honor has been soiled, and I find myself consoling him, assuring him that *no hay que preocuparse*, we've been in worse situations.

That night, under the long twilight aura of the Canadian sky —as far north as I have ever been, as far from Chile as I have ever been—a council of elders is convened, my parents and Angélica and me, and the group quickly reaches a consensus. We are not embarking on another adventure, and certainly not to the labyrinth of Mexico, where the rule of law is arbitrary and the most powerful man in the land is a sworn enemy of my protector. We do not particularly wish to dwell in the United States, but it would seem that, for the moment, there's no other option. We can't go back to Europe, we can't travel to dictatorial Argentina, Chile is out of the question, and at least we're semi-installed in Washington, and my English, yes, my very own neglected English, will help me make a living of some sort now that we're stranded.

The next morning, I call up the Fentons, who own our rental in Bethesda, and good news: they were about to sign up new tenants, but they like us, especially our Joaquín, who has charmed them as he does everyone into whose life he toddles and tumbles — so we can stay an extra year. We also manage to get Rodrigo registered again at his French school; my parents will take care of the fees now that we are abruptly repauperized. Rodrigo is happy to avoid one more uprooting. And Angélica confesses that she feels a bit guilty, even responsible for the fiasco. She'd been apprehensive about Mexico, our fifth major country in eight years, that capital city overrun with crime and pollution, and had foreclosed the possibility, she said, by resorting to one of the superstitious rituals of the Andean countryside where she grew up: tying a knot in a handkerchief, called a *pilatos,* and beating it unremittingly and threatening more reprisals if her wish to remain in the United States was not granted.

The knot was untied and spared further carnage, but . . . be careful what you wish for. Our J-1 visitor's visa was about to expire, I had no job and no legal permission to search for one, no health care, no Social Security, no safety net, no independent income, no credit card, and no credit, for that matter.

Slowly, we overcame each obstacle. Our dear friend Isabel Letelier — still living in Washington after the murder of her husband, Orlando, by Pinochet's thugs — recommended Michael Maggio, an immigration lawyer active in the solidarity community, to help us out. His tactic to secure a change in our resident status was an original one. Most people argue that they want to stay in the United States permanently, but Mr. Dorfman does not wish to do so, Maggio earnestly explained to the INS agent. Mr. Dorfman will gladly return to Chile if the United States can induce the Pinochet government to accept him. So you won't be remaining here, Mr. Dorfman? No, sir, I'm going back as soon as things change in my own country. The officer was nonplussed enough, perhaps moved by our plight, to consider granting us a green card, and further influenced by offers of employment from the Institute for Policy Studies in Washington. It was not clear to my lefty friends at IPS — Mark Raskin, Dick Barnett, Saul Landau, Bob Borosage — whether there would be funds for the post they had advertised — which they actually opened exclusively for me (it turned out there wouldn't be a penny), but without that letter of intent I would have been in real trouble.

I may have been helped by the coat I was wearing. Maybe that was what impressed the INS agent, that mark of my status.

My coat?

I still own it today. It hangs in a closet of our house in North Carolina, on a rack crammed with half-hidden clothes, a smooth, honey-colored camel's hair coat of wonders that became, over and over, my mysterious shield against the insults of the world.

I hardly knew the man who bequeathed it to me. Draguy Nicolitch was a gaunt, handsome, elderly Yugoslav who had spent most of his adult life in France. We had met one afternoon by chance at a hotel in Brela, on the Dalmatian coast, during a 1975 summer holiday — funded, as usual, by my parents — and had taken an instant liking to each other. He invited our family over to the table where he was dining with his Parisian wife, Henriette, and proceeded to recite Neruda with a French accent barely more comprehensible than my own. I noticed how his fists clenched when I told him how much I missed Neruda's grapes, the taste of my homeland, and the sea, as blue as the one washing Draguy's own native shores. Later in the holiday, I learned from Henriette that he had been a Maquisard during the Nazi occupation and that they had long been cooling a bottle of champagne in anticipation of Franco's demise, and, with a sigh, they'd added another one for Pinochet on the day of the Chilean coup.

We agreed to meet again in Paris. Sure enough, when the telephone rang one evening late in 1975, it was Henriette. A few days after our departure from Brela, she said, Draguy had walked into the sea and died of a heart attack, just like that. I could imagine the tall, magnificent old man obstinately stepping into those Adriatic waters as if he had a rendezvous with the memory of his childhood and not with death. "I have something for you," the widow added.

It was the sumptuous coat.

Henriette told me that several times after we had left the resort in Yugoslavia, Draguy had insisted, prophetically, that if he died, it was to be mine. When she brought it over, I thanked her without revealing that I'd never wear anything that prim, that expensive. Back in Chile, Angélica had always bought my clothes, going off to a shop, choosing something, and returning later with me to try it on, a situation that persists today, since I still don't know my measurements, my neck, waist, or shoe size. I'd venture forth with different-colored socks if my wife hadn't trained me over so many eons of marriage. At any rate, in Paris, with our borrowed, secondhand existence teetering on the stark edge of survival, Draguy's shining apparel, fit for a prince, felt particularly out of place.

I didn't, in fact, put on the coat until a year or so had passed and we'd reached Amsterdam. And then one day, just to remember Draguy, I draped it over my lanky Don Quixote figure.

That was no exile in the mirror. That was a gentleman.

The old Yugoslav had migrated to France when he was young and had learned early on what it means to be despised because of your language, discriminated against because of your lack of papers. He knew what it was like to stand in line with other immigrants and how the police hound the undocumented and how fascists loathe those who are different. Perhaps it had been during one of our conversations, when I had admitted the difficulty of persuading people to help when you've been deprived of the subtlety of your original tongue, when your clothes are drab and worn, perhaps it was then that he decided to endow me with his coat of arms, understanding that it would defend me from more dangerous threats than winter.

I wore it to my first session with the Dutch immigration authorities. A mere formality in my case, but still four or five hours of dread, made more stifling by the overheated waiting room filled with petitioners, job seekers, refugees, illegals, Turks, Moroccans, Slavs, Pakistanis, Mexicans, Indonesians, Guatemalans, Nigerians, every nationality, every creed, every color, fearful that something will go awry, waiting, waiting even when they leave with a stamped piece of paper, only a reprieve as ahead looms the next waiting room manned by capricious bureaucrats. Waiting rooms of Babel, like the one in Buenos Aires that I had sat in for days hoping to be allowed to depart Argentina before the death squads came for me. Or the one on the Île de la Cité in Paris where I hoped they would allow me to stay, and the waiting room lurking in our future in Baltimore when we had run aground in the States. So many similar rooms, each one disgorging the remnants of yet other rooms in faraway countries, so many different skins and pitches of voice and beaches of tragedy, those mothers suckling their babies, those fathers slapping the rowdy kids, warning the oldest ones not to touch, not raise their voices.

Fleeing from lands mired in calamity, immigrant cooks, garbage collectors, runaway maids, students desperate not to be deported—those rooms where I came into ephemeral contact with the forgotten flotsam of the world.

The rounds in Holland were more arduous than in my other countries of exile. The police would demand a work permit first, and the Labor Ministry insisted that no, first I had to be a resident, a certificate was missing, an x-ray of my lungs was blurred—what if you have tuberculosis? A document from back home needed to be notarized, the right signature in the wrong place—or was it the wrong signature in the right place? The University of Amsterdam finally sorted things out, but only because my contract was to expire at the end of four years. Why four years? To make sure you couldn't remain in Holland. (After five years one had the right to apply for permanent residency.) I didn't mind. I expected Pinochet to be overthrown tomorrow.

Meanwhile, my coat worked miracles, at least insulated me from the quicksand emptiness in my stomach. And impressed the officials in Holland, as it did in the United States. Doors opened, smiles bloomed, patience was bountiful.

The coat came out many more times, with a tie to match, thanks to ever-vigilant Angélica. I put it on each year to renew my visa, when I had to ask a favor for Chile, when I accompanied our comrade Enrique Correa to meet some minister or important contact, when I had to face the tax inspector, or when I needed a loan at the bank. I put it on to defend Rodrigo when his teachers accused him of being unruly. Oh how I wish I had donned it that day in Paris when I had been hauled in by the principal at the school in Vincennes. He was an adroit, wiry Frenchman, punctiliously courteous, but sharp as a razor when he told me that my son was behaving abominably, that he refused to eat the cheese at the end of his free meal, *mon Dieu, mon cher monsieur, pourquoi est-ce qu'il ne mange pas son fromage, pourquoi cet enfant est tellement difficile?* Rodrigo was such a pain in the ass that instead of scarfing down his Camembert he'd tell the teacher watching over him: You're also missing your recess, so I don't care, I can stay here the whole hour. The principal was ever more incensed, *mais, c'est pas possible.* Noticing a photograph of Mitterrand and a copy of the socialist paper *Le Matin de Paris* on his desk, I was inspired to explain that this was a cultural problem, Chilean boys don't eat that kind of cheese back home. It was cruel to force this exiled child who had lost everything and had nightmares about Pinochet and the teachers who had been massacred, it was unfair and lacking the milk of human kindness (believe me, I was not this eloquent, I did not quote *Macbeth* — and certainly not in French!) to keep the lad eternally in front of a piece of redolent Roquefort — but the principal was all solidarity with Chile, yet felt none for the specific Chilean kid who refused to do in France as the French do, unwilling to become Asterix or Obelix, a true Gaulois. And all the while, back in our apartment in Vincennes, I had the key to assuaging this guy's vanity and stirring his admiration: I should have shown up under the mantle of Draguy's coat and all would have been *bien, tellement parfait.*

The coat, the coat! Protecting me when I divined trouble crossing frontiers, and once, just once, worn in order to rescue our own Antonio Skármeta, who had been detained when he came to visit us in Amsterdam and got stopped at the German border because something was wrong with the seal on his passport. I called Mineke Schipper, a friend at Dutch PEN, and she contacted the Ministry of Foreign Affairs and said she'd get back to me, but I didn't wait — my friend was being held by the Dutch police, had somehow convinced them to take him to the Buitenveldert precinct, explained that his eminent friend Dr. Ariel Dorfman would bail him out. You should have seen the cops' eyes when I entered with my coat. Among the drunks and petty thieves and hooligans and a woman who looked like a hooker with a bruise on her lip and a black eye, I was simply resplendent. Like a medieval knight in search of the Holy Grail bearing a potion against wounds, I understood why warriors stood guard over their coat of arms on the night before a battle, why they invoked invulnerability as the day dawned.

Not the only acts of magic the coat accomplished.

One day in 1978, just before Christmas, I received a package from my mother-in-law. She'd been sending me press material from Chile on a weekly basis for years, articles and opinion pages diligently perused, along with listings for car and house sales, employment ads, obituaries and wedding announcements, anything that would give me the illusion that I was back there. Of late, Elba had added *Revista Hoy,* a weekly magazine created by the opposition that was continually on the verge of being closed down. It was run by Emilio Philippi, a Christian Democrat with whom my relations had not always been cordial. He had been the head of the Union of Journalists back in 1967, when I had resigned with great fanfare from *Ercilla* because its owner, a man named Torreti, had refused to print my interview with the eminent Cuban poet Nicolás Guillén. Apparently, Torreti had exclaimed, upon reading the proofs, "I'm not publishing any nigger Communist in my magazine." When Philippi had reacted in lukewarm fashion to that flagrant act of censorship, I had denounced his lack of support and encouraged my two hundred students at the university's journalism school to protest outside his office.

A decade later, however, I found myself sending that very same "cowardly" Philippi a letter containing an article about Draguy's coat. I fully expected him to ignore me, not out of hostility but simply because, as far as I could tell, my name was *verboten* in Chile. I would have loved it if the right-wing papers in Santiago had thundered against me, reviled me, published exposés of my anti-junta activities, but nothing, only silence, as if my existence didn't register on their radar.

"That's why there's no answer from *Hoy*," I told Angélica. "They wouldn't dare publish me. I'm too dangerous." Angélica shrugged; she didn't think the article was especially inflammatory. And could I help her with her breathing exercises? She was enormously pregnant with Joaquín and there were more urgent matters to deal with.

Then a package arrived. And there they were, in *Revista Hoy*, my words about the coat! Not an adjective, not a verb, not a noun missing! How the coat shielded me from more than ice and wind, how Draguy, that Yugoslav fighter against fascism, had understood why someone like me might need a magical cloak. The story of our exile and penury, our wanderings, disgrace, dignity, all of it circulating in Chile. Directly, from that page and into the eyes of those who lived in my country. People I knew and people I didn't know had lifted their eyes from my words and seen, not the gray winter skies above Buitenveldert, but the blue summer of an imperfect Chilean day. They did not have to hear, as I did, the cries of children in a language I could not decipher, the eyes of my readers in Santiago returned to the article and alighted on a photograph of me accompanying the text, rescued from who knows what forlorn archive. I looked as if I were sixteen, adding a waif-like, orphaned susceptibility to the piece itself.

Draguy's coat, then, had taken me home. For a few minutes, as I read my own words under the strange, unfamiliar light of Holland, I found myself transported back to the land forbidden to me.

A fairy tale. A dying man gives a young wanderer an amulet to protect him from the evils he will encounter in his travels.

Seen, however, from the perspective of thirty years later, I can discern that I invested that mantle bequeathed by Draguy with magical powers during a period when, no longer in dire shape, I was making some of the choices that would gradually reinstate me among the powerful. If anything, it was a transitional object that I could have discarded without any major upheavals in my upward trajectory, but that allowed me at a time of psychological distress to cling to something safe each time I felt myself sinking, when I doubted whether I would rise again and survive exile.

But, of course, the real coat I needed to help me survive was the English with which I forge these words, the real shield I was to wield in the country of dragons where we had been stranded, that I was to carry back with me to the Chile brimming with monsters of another sort.

As if English could protect me from what awaited me there.

Fragment from the Diary of My Return to Chile in 1990

SEPTEMBER 1

I have spent the last few weeks standing in line.

Ever since I finished cleaning up my library and sent young

Miguel off to learn the profession of bricklayer, my energies have been absorbed elsewhere.

I happen to be an *exonerado*, one of thousands upon thousands expelled from their jobs right after the 1973 coup. The new democratic government is offering people like me the chance to be considered for reinstatement to their former positions.

I'm less ardent about this prospect than a number of my former colleagues who have returned here with doctorates and a list of impressive scholarly publications and are wearily jumping through an array of administrative hoops. What right does the current faculty of the Spanish department, most of them mediocrities who stayed on after the military "intervened" the university (meaning they installed an admiral as rector/provost who cleaned its halls of "subversives"), what right does that academic detritus have to judge where and whether I fit, by what authority? Why am I not judging them? Why don't they resign, have everyone compete for reinstatement? But in this, as in so many other cases, those who benefited from the dictatorship are untouchable.

So no, I don't think I'll be teaching in Chile in the near future, can't see myself smiling day in and day out at those tawdry, tedious academics who smiled when I was being persecuted. I wouldn't mind regaining the pension funds purloined by the dictatorship — ten years of Social Security deposits, first as an assistant, then as an associate, and finally as a full professor of Latin American literature — maybe even an apology. Though what really fascinates me is the possibility of discovering how my discharge was justified. I'd love to read the files stowed away in some dusty vault that spell out why I was forced out.

And so, on and off for the past few weeks, I've been repeating here, in my own country, the dreary waiting rooms of exile. At the end of days of slogging from one office of the University of Chile to the next, one dilapidated building after the other, and being given the habitual runaround — could you notarize this document, this facsimile can't be read clearly, and why does it say Ariel here when your name is Vladimiro Ariel? — the information was handed to me. I was ousted, the official decree states, because I was subverting the young and professing totalitarian ideas. There is a revelation I did not expect. Some anonymous official — not content with stealing my pension — had illegally reinstated me in my job for eight months, so he could pocket my salary while I was wandering penniless in Buenos Aires and Paris.

It is that petty pen pusher and people like him who kept Pinochet's regime afloat all these years, they have been the silent accomplices thriving in the shadows, taking advantage of our tragedy. How many are out there, the leeches, the parasites, the ones who cut a deal, the ones who have a stake in suppressing their pedestrian roles in the repression? That man — who will, of course, never be held accountable — did not, however, ruin my life, something that most of the men and women who sat on benches with me cannot say. That's what I really learned from the days filling in forms and waiting around for some somber functionary to attend my case. There's the hospital worker who hasn't had a job since he was ejected years ago, and his daughter who was kicked out even though she was uninterested in politics, "just because we shared the same name." And the janitor from biological sciences with a beret like Neruda's, briskly trying to belie the cudgeled expression of the face below. It's as if the old man had been *apaleado*, beaten with a stick. What makes it all the more heartbreaking is their stoicism, how they refuse to complain.

Nevertheless, I sense that something more than dignity is involved in their reticence to elaborate on the lives they have led. Fear — it is the same pervasive dread that has turned so many faces into impassive masks, behind which real thoughts and deep feelings are carefully concealed.

It was one of the first things I realized when I was allowed back seven years ago, in 1983. *Ud. no habla en Chileno,* over and over that's how dissidents described me, they insisted with averted eyes that I don't speak like Chileans do — literally, don't speak "in Chilean." Meaning: I had not become skilled at dissimulating, my voice was too loud, I somehow still inhabited the safe space of expatriation, I was trying too hard to speak across the abyss separating me from those who had lived through the years of terror. Insisting that they should tell their stories, that this was the only way to heal.

Is it really?

A week ago, I ran into a MAPU comrade who, like me, has just come back from exile, and he told me how, after the first dinner with his right-wing family, his mother noticed that he was dragging his left foot slightly as he shuffled towards the living room. "What happened to you, *hijo?*" she asked. "Did you hurt yourself?"

"You know perfectly well why I'm limping, *Mamá.* I was tortured, that's why. I'll never walk normally again, you know that."

Tortured? His mother looked at the other members of the family as if to excuse the wayward child and his pranks. Of course the boy hadn't been tortured, *hasta cuándo* was he going to engage in that sort of political propaganda, let's not dwell on such unpleasant topics, we're here to celebrate your return, *mi amor.*

My comrade thinks that his mother has truly forgotten what happened to him, decided that she couldn't live with herself or be at peace with the Chile of her beloved General Pinochet if she admitted her son's agony. How to break down that barrier? Did it make any sense to thrust his injured leg and humiliating story into every conversation, to vomit it up as the chilled Chardonnay was being poured into the wine glasses and the chitchat started? Could he live like that, constantly acting out his traumatic experience for an audience that did not want to know, did not seem to care?

"I'll just keep quiet," he said to me. "What else can I do? Shout from the rooftops what I went through, create a scandal whenever I open my mouth?"

I'm not sure what the answer might be, have not really known since I arrived here on that 1983 visit after ten years of exile.

The whole family was gathered to welcome us at a luncheon banquet at my mother-in-law's — and since they're all Allendistas, there was no need to lie about what happened to any of us. But the merriment was more than I could handle, so I murmured an unobtrusive excuse and slipped out the door, went down a shaky elevator and onto Monjitas, a bustling street in the center of Santiago.

It was a reprieve to be on that corner, merely to watch the traffic go by. It had been overwhelming, all the friends and relatives who came to the airport with flags and chants, hugs and gifts and tears, and I needed to soak in the city by myself, steal some moments alone, intoxicate my lungs with blessed gasoline fumes, overhear two workers go by telling a ribald joke, observe the slender women, the sauntering women, the broadassed women, listen to the stiletto heels of their shoes clack with a rhythm you can find nowhere else — and that's when I saw the tree.

A stunted tree, skinny, dry, leafless, planted in a patch of soil crowded by pavement, one among myriad other puny trees barely subsisting on the sidewalks of Santiago. Somebody had tried to help it grow straight by tying its scrawny trunklet to a stick with a torn rag.

I had tried to hang on to every last detail of the country in my mind, intentionally ignoring that my endeavor was useless, each memory a brief, sharply focused stab of light in the darkness beyond whose murky edges a vast extension of experience was irretrievable. Whenever I submitted to the trap of nostalgia, I ended up subtracting more than I added. I collected photographs, persisted in old habits, asked visiting friends to bring herbs from home. And yet, in my fierce quest against oblivion, I had not once recalled that tree in all its slight, modest dignity. Not once in all my years of banishment.

Nevertheless, this failure to evoke its irrevocable perseverance did not trouble me. On the contrary, the sight of it flooded me with relief.

This rickety tree told me that I was not, as I feared, deracinated or shipwrecked without an identity, that I was then, as I am now in 1990, in love with this land and that it loves me back. It assured me, that tree, in all its splendor and glory, that I had never really left, that I'd been gone, if at all, for a few minutes, this frail, abandoned messenger derided the fear that absence would turn me into a stranger. It was as if, for an instant, those years in exile had been abolished, as if I could step outside history.

The roar of an engine broke the spell. A truckload of soldiers passed by, awakening me to where I was. Not twenty blocks from where the tree and I communed was the real master of every shrub and every leaf and everybody who might look at it, the one who could make me disappear with the flourish of his finger. That tree was a captive in a captive land and would not protest anything that happened to me or anybody else; that's how it had survived. And somehow it was all right that the tree, precisely because it was neutral and untainted, also belonged to my adversaries, to those soldiers in the back of the truck that was idling at a red light, to that con-

script peering out from behind a tarpaulin, his eyes hardly visible under his helmet, but black and intense, probing, hostile,
expecting an attack. For a second I assumed he was looking at
me, but no, someone else was in his sights, a man stood near
me, a few feet away, examining the truck. I had been unaware
of his company, but my sudden interest must have perturbed
him, because the man abruptly turned and, however briefly,
our eyes met.

I could not read those eyes of his, tell what he thought of
the military truck as it revved its engine, what was behind that
young soldier's cold stare as it disappeared in a belch of smoke.
The last time I had walked these streets, in democratic times, I
could have separated friend from foe, I could have deciphered
the constellations of anxiety or hope behind those eyes.

Now, nothing.

That man had learned to hide.

Seven years later, I still have no way of telling if someone
like him hates those soldiers or welcomes them or is as unconcerned as the tree, no way of knowing if he is the sort of man
who would tie a rag to an ailing plant to help it grow or hack
it down if it stood in his way. I have no way of understanding yet, not on that return, not on this return in 1990, what has
been done to his soul, the lost soul of this country.

Perhaps, after all, nothing can bring back the hour of splendor in the grass, of glory in the flower.

Tal vez el duelo por mi país solo acaba de comenzar.

Perhaps the grief for my country has just begun.

•

SPANISH DID not give up its privileged position in my life just
because we were shipwrecked in the United States.

It fought tooth and nail and throat to make sure it was the
only one to keep us afloat. While I finished my *La Última Canción de Manuel Sendero* — about babies even more *desamparados*

than we were because they'd called a strike against adults, refused to be born until the world had been radically reformed—I eked out a living teaching classes in Spanish, jobs concocted by caring friends at nearby universities; writing a report, also in Spanish, for UNESCO on theatrical arts in Latin America; and, more lucratively, writing articles, two stories a month for *Triunfo* in Madrid, *Proceso* in Mexico, *El Nacional* in Caracas, and any other periodical that I could wrangle a fee from.

My Spanish was bolstered by the unforeseen circumstance that the United States was not a monolithic Gringolandia. The Mexican president may not have wanted me to live among his people, but millions of his unfortunate compatriots were trudging through the desert, squirming under barbed wire, coughing through fetid tunnels, being betrayed by coyotes, and running from the Border Patrol, all to come to *los Estados Unidos de América, el Norte*, where I had been stranded in a city, Washington, that, in fact, was trending Latino. Well, areas of it.

Above all the Adams Morgan neighborhood, where Angélica worked counseling women, most of them expecting babies. Some had been raped or abandoned on their journey north, others were being abused by their menfolk, and it was my wife's job to guide them to adequate health care and, once the child was born, to ensure that formula and food stamps and medical attention were available. With her ever-generous heart she accompanied them above and beyond the call of *el deber*. She ended up being present at the births of their babies, visiting their houses, scolding their husbands, and finding clothes for the little ones, all of which required an immersion in the culture and customs of her *protegidas*.

Once in a while I would visit the building where she worked, the Family Place. So delighted to be back on the continent momentarily denied to me, braced by a neighborhood almost entirely taken over by recent Latino migrants. Just to linger on the streets and hear Spanish spiced in all those accents and rolled off tongues that had tasted the waters of the Caribbean and the Golfo de Fon-

seca and the Lago de Managua, to listen to the boleros and cor-
ridos blaring out of stores, to see the Safeway on Columbia Road
packed with imported produce and labels in Spanish — all this
hinted at a potentially bilingual America.

These were not the reliable categories into which I had segregated geography: if you speak Spanish you live down there *en el
Sur,* if you speak English you live up here in *el Norte.* Not ready,
when I first arrived in the America of Ronald Reagan, to see myself yet as part of a vaster social movement, a seismic migratory
shift within which I could be defined as a literary wetback or an al-
ter-Latino, a term I have recently coined so I can sprinkle it liber-
ally in conversations. I couldn't have predicted that I would someday live in North Carolina, the state that boasts the fastest-growing
Latino community in the United States. Or anticipate that Rodrigo, with whom I had labored to keep Spanish intact, would, as
an adult, use that language of his birth not primarily in Santiago,
but in Durham, of all places, in order to create daily links with
Mexicans and Venezuelans and Colombians and Ticas and Nicas
and every other South of the Border nationality, making a new
life for themselves in the New South of the United States. How
could I imagine that Rodrigo would feed his family by chroni-
cling in his films the troubles and dreams of those migrants? And
how to even conjure up a day when I'd be asked, as I was only a
few months ago, to pen an anthem for the Cooperativa, a local
community credit union founded by Latinos for Latinos. I took a
popular *corrido mejicano* that bemoaned the fate of a lover destined to lose his woman and trying to get her back, and adapted its
words to urge the Cooperativa's clients and participants towards
autonomy and economic self-sufficiency and yes, pride, at a time
when anti-immigrant fervor was reaching a feverish, nativist, Tea
Party pitch.

Nothing could have been further from my mind in the early
1980s.

My aspiration was to live in monolingual Chile, rush off as

soon as I could to the same South forsaken by all those pregnant women cared for by my wife. I did not discern my dip in the Hispanic waters of Adams Morgan as a voyage to a gloriously multicultural America of the future, but rather as a refreshing connection to linguistic allies from the past, just when my resurgent English had started to perilously lure me back into the fold.

Because a catastrophe — for once not descending directly on us — had opened an unexpected door to a first flirtation with my childhood sweetheart, the sweet syllables of an English I had up to that point in my wanderings only scribbled unwillingly in the prosaic service of our cause.

The year 1982 opened with a snowstorm of such fury that Air Florida flight 90 crashed into Washington's 14th Street Bridge, killing seventy-eight people but also inciting bystanders to acts of exceptional nobility, saving survivors from the icy waters of the Potomac. The next morning, as I looked out from our Bethesda home onto that white wasteland and the devastated, paralyzed city beyond it, something stirred in my mind. Something in English, not Spanish. At first remembering the violent winter of 1948 when, as a child in Queens, I walked to the curb through walls of snow taller than my own pint size. But that's not what was really swirling inside.

After eight years of exile from Chile all I have to do to feel at home is look out the window. To a Latin American living in the States, this winter has turned the scenery into something sadly familiar.

Inspired, scrabbling up words from a well of emotion, I let English invade me, possess me: *It's not the snow. We hardly have any of that down south. After all, we must live up to our caricature of taking siestas to escape the sun.*

And more: *It's the disaster that is familiar. All I have to do is look out the window and deep into the TV news, and there I am, back again, as if I had never had to leave my country. Millions of children aren't going to school, people are dying on the streets, transportation is a mess, countless homes have no water or electricity, factories are functioning under capacity. There's plenty of food in the stores, but you can't get it. And the rhythm of life has lost its usual briskness. Nothing arrives on schedule, as if we had been caught in a slow-motion — or is it a snow-motion? — camera.*

Puns? Jokes? Playing with the language like a native?

I *was* a native, that's what the rhythm of my words and the rhythm of the landscape revealed, that I loved this language banished from the carnal circle of my existence. I loved every word I had learned as a child and that I could no longer disinherit as I wrote that our snow down south was invisible, *if our perpetual snow were to bear names, as hurricanes do, it would be the sorry names of dictators, one mask after another on the same face.* And on and on, as if I were a dark blizzard, because that was what I was really writing about, that our invisible snow in the South is man-made and would not melt away in the spring. It had become such a habitual part of the territory that we tend to forget that we are living in a state of permanent emergency that should bring out the best in everyone, as we do when a cataclysm hits us and turns societies, I wrote, *into brother- or, if you will, sisterhoods, into gigantic tribes or families.*

And I flew to the end of what I was voicing: *Maybe we need a daily misery report just after the weather report, inform people each plentiful morning how many inches of suffering still remain on this earth, how many more are to descend before nightfall . . . Because I continue to believe that the winter called poverty, ignorance, injustice, repression, could be melted by our common concern.*

There it was, finished. But — what was *it*? This enigma had stormed out of me and now demanded to reach others and share with them the healing of the world but also of my divided heart and mind. It had to be published, but where, what to do next? Maybe my friends at IPS had contacts at a newspaper?

A few hours later, the telephone rang in our house in Bethesda.

"This is Howard Goldberg. I work at the op-ed page of the *New York Times* and we'd like to publish your piece. Tomorrow."

A few years earlier, back in Holland, I had been filled with joy upon discovering that my Spanish words in *Hoy* magazine had guided me home in spite of the Chilean censors. And now a mirror moment arrived in English, my return to a place I am only tentatively willing to call home, even now, almost thirty years later as I write this memoir in English, even after I have selected English as the tongue in which to puzzle out my identity, even now I tremble as I state that a parallel jubilation to that experience in Holland was blowing through me like the wind outside, like the wind of words inside welcoming me as I welcomed them, January 15, 1982, there it was, my commentary in the most important newspaper in the world. And for that ego tattered by exclusion and baffled by the confusion of a mongrel identity, for that man with no job and no government to protect him and his family, there was a certain vindication: so, General Pinochet, you sent me to die in the wasteland of exile, and look what I did with your death certificate, look what I turned your hatred into!

But that inaugural publication of mine in English was more than a one-time exploit. It also established the terms of how I would function as a public intellectual in the years ahead. Stranded in the land I had loved as a child and then had deplored as an adult, gentle and hostile to it, angry and sympathetic, I was being afforded a chance to see this democratic empire up close, to survey it with my now Chilean eyes. That attempt in the *New York Times* to reach out to readers in the States was the first of many explorations of the country where I was marooned from a perspective and field of feeling that I had perfected throughout a Latin American existence. By embarking on that path, however, I was also inviting myself to do something similar with Latin America itself: gaze at the continent of my origin with fresh eyes, my now distant eyes, my suddenly strange eyes, abandon the pretense that

I could write about Chile from the States in the same way I would have written if I had never left.

That morning in mid-January of 1982, the opinion page of the *Times* reminded me that I was as divided as the two snows that plagued my life: outside and freezing in the streets of Maryland, and staring back at me invisibly from inside the mind of summer of Chile. As divided as my two libraries, it was then, I think, that I began to inadvertently walk on the path to the person I am now, someone able to write these pages here in Durham, a bridge spanning cultures, someone who has turned the curse and blessing of being both a stranger and a native in the United States, the blessing and curse of being a native Latin American and also an alien, into a source of persistent, fruitful tension.

It was just a start, a mere hint of the work of reconciliation that loomed ahead. To make a friendly overture to English after years of enmity forced me to look deep into my homelessness and simultaneously embrace that condition, celebrate what it revealed.

A rehearsal for the memoir that I now write, in this country where I never intended to stay. A glimpse through a half-open portal offering a different way of traveling to Chile, the possibility of using that country where I yearned to live then and do not live now as a prism through which to view the contemporary world.

Something as ephemeral as one snowflake, and then another one, and then one more, accomplishing so much: anchoring me to the urgent immediacy of the United States and simultaneously securing me to Chile through metaphors that forge allegiances that are just as powerful as geographic location, so much hidden and exposed in that blizzard, because it also promised, if only I had been able to listen, that the land I would always inhabit was the land born of the compassionate imagination, the land from which only death can ever dislodge me.

PART III

DEPARTURES

[Warren:] 'Home,' he mocked gently.

[Mary:] 'Yes, what else but home?
   It all depends on what you mean by home . . .'

[W:] 'Home is the place where, when you have to go there,
   they have to take you in.'

[M:] 'I should have called it
   Something you somehow haven't to deserve.'

— Robert Frost, "The Death of the Hired Hand"

THERE IS A carpenter in Chile, a man called Carlos, who can teach us something about fear.

I met him while filming Peter Raymont's documentary in 2006, when I spent a few hours in one of Santiago's poor neighborhoods, distributing books from my library to residents for Libros Libres, an unpretentious do-it-yourself organization created by Raquel Azócar, a woman with whom I'd been collaborating for years to "liberate" books into different communities. I had just finished reading one of my stories for children to some of the neighborhood rascals when I was approached by an old man. He'd heard that I'd worked with Salvador Allende and had something to tell me.

Carlos had been an enthusiastic supporter of Allende, basically because his government had created a program that had helped that carpenter to purchase his one and only house. After the military takeover, when soldiers raided his *población,* Carlos had been so terror-stricken that he'd hidden a picture of the martyred president behind a wall of his house, where it remained throughout the seventeen years of the dictatorship. He did not retrieve it, Carlos informed me, even when democracy returned to Chile and Pinochet had to relinquish his stranglehold over the government. Nor were a series of free elections enough to release that carpenter from his dread. It was only in 1998, when General Pinochet was arrested in London for crimes against humanity, that Carlos pried back the boards that concealed the portrait — and there it was, after twenty-five years, his *Presidente lindo,* his beautiful president, he said, just as he recalled the man. And when Pinochet was flown back to Chile after eighteen months of London house arrest, Carlos gathered his courage and kept the picture of Allende hanging defiantly on the wall. Never again, he said, was he going to conceal it.

It's an inspiring story, because Carlos was not a militant, a soldier of the revolution sacrificing himself for the common good. If that portrait from the past could emerge from its hideout and share Chile's air and mountains and grandchildren, it was because Carlos had refused to forget; he had not burned the picture while the security forces rampaged outside but had buried it furtively both inside his mind and behind a wall until it could be recovered.

An inspiring story, yes, but also sobering.

Memory, like courage, does not exist in a vacuum. If there had been no justice, if Pinochet had not been made to face judges and answer for his crimes during that year and a half in London, that portrait would have remained in hiding. For the memory to flow out into the open, the fear also had to flow out, there had to be a societal space where the portrait from the past could be safe. The surfacing of those proscribed images and thoughts had itself been the product of many other, more communal acts. Carlos was eventually able to bring together his private and his public memory because others risked everything so a commons of liberation might exist. The case of Carlos the carpenter is sobering, no matter how fervently admirable his loyalty, because the very isolation and secrecy of his hideaway also reveals how ultimately precarious any merely inner and covert rebellion can be, and what happens to a country like Chile when its citizens decide to let fear, even during a transition to democracy, rule their lives.

In Santiago, we employ a handyman whom I will call Rolando. And though he'd had dozens of conversations with Angélica while he painted, tinkered with the plumbing, and sanded some doors, it was only after Pinochet was indicted publicly in a London court in 1998, the day after prisoner Pinochet had been photographed in the dock, that Maestro Rolando had, for the first time since he began working for us, revealed to my wife the most traumatic experience of his life.

A few years after the coup, he said in a matter-of-fact voice, he'd been arrested and tortured by General Pinochet's police. Rolando had worked back then as a porter in a school, and his tormentors had wanted him to implicate his colleagues, to finger teachers who might be engaged in "terrorism." It had been a brief detention. Two, three days and then they let him go. As a result, he lost his job, suffered bodily pain for a few months, and psychological damage for who knows how long.

For almost twenty years, like Carlos the carpenter and millions of compatriots, Rolando had shut himself inside the closet of his secret emotions, had murmured the tale only to his own inner shadow. What had freed his voice, given him the courage to tell his story to someone like Angélica, was the certainty that Pinochet was not above the law. That's what had done it, the six English policemen who escorted the former dictator into a courtroom in London to face his accusers, the fact that Pinochet's lawyer had asked Magistrate Parkinson to grant his client permission to walk in the garden. There it was, the proof of the General's vulnerability and decline: if he wanted to walk in the garden, he had to ask permission!

"And you're no longer afraid?" Angélica asked Rolando.

"Only a fool would not be afraid."

I know what he meant. I'm not immune from that fear, I have tasted its bile and the far worse bitterness it leaves behind, how it prevails beyond the moment of violence, how hard it is to be rid of the serpents of apprehension once they have snaked their way into your mind.

On that trip to Chile in 2006, I visited the Salvador Allende Foundation, set up to commemorate our dead president's memory. In spite of being a member of the foundation's board, I had not yet inspected the recently acquired house where the archives and museum were located, and wasn't aware of that sprawling mansion's peculiar history. So I welcomed the invitation by Patty Espejo, an old classmate from university days who had gone on to become one of Allende's most trusted secretaries, to tour the residence with her.

"We didn't know it when we bought this house," Patty said as she guided me and our camera crew down to the building's basement, "but it belonged to the DINA — Pinochet's secret police had one of its headquarters here. Take a look at what they forgot to take with them. On purpose."

I had written extensively about the invasion of our private lives during the dictatorship, but even so was not ready for that underground cavern, those lower depths where Pinochet's Gestapo had spied on Chileans, leaving behind a warp of twisted wires splayed in a multitude of bright colors, which made them all the more perverse. The agents — nameless, unscathed, unidentified — had not dismantled that tangled snarl of wires, in order to flaunt a message of impunity, to sicken us with the warning that they had heard us talking to each other, had listened to our secrets, that the power they once had still endured.

I had been one of those captured inside those listening devices.

On my first trip back to Chile in 1983 I had jauntily agreed to write an op-ed for the *New York Times,* to be published on Sunday, September 11, the tenth anniversary of the coup. Howard Goldberg, my editor at the paper, had instructed me to phone New York and dictate my commentary days in advance so he could contribute editing suggestions.

I started my piece by stating how normal everything seemed, much too normal, that I had somehow expected the birds to sing differently under a dictator. While everyday life, the taste of the food and the way people laughed, remained unperturbed, my ten years away had made me acutely aware of the abysmal distance between rich and poor that had grown malignantly during the General's forced modernization. Nothing I had read about changes in Chile had prepared me for what I felt when, after passing through a virtually unaltered Santiago full of misery, I reached the *barrio alto*, the hillside neighborhood where the privileged classes traditionally reside. I found myself being guided, like a tourist, along unknown avenues filled with hundreds of glass towers and shopping malls, splendid gardens and efficient freeways. Only a few miles from that sleek and exclusive city-inside-a-city were the slums where millions of impoverished Chileans lived in squalor.

"It is as if Chile," I said into the telephone, "has been struck by a plague."

Since midnight I'd been feverishly scribbling this piece, determined to prove to myself, perhaps more than to others, that I would not let any adjustments or inhibitions worm into my writing just because I was back home. I'd try to be as impudent in perilous Santiago as I had been abroad, where no reprisals could reach me.

Sixteen-year-old Rodrigo wandered into the room and sat himself down, allowing a mixture of admiration and alarm to creep into his eyes, finally whispering, "Hey, man. That's strong stuff."

I threw a pillow at him and he dodged it, then sat there listening to the stronger stuff that was on its way, my accusations of censorship and torture and murder. As I spoke these words, I began to hear my voice through Rodrigo's ears, a reminder that I was spouting this indictment in Santiago, where a secret police agent was probably noting each phrase in a cellar not far from where I was sitting. But I calmed my beating heart in 1983 by telling myself that I was protected by the international press, that CBS had sent a team to film this first return of mine, that Pinochet could not be so stupid as to arrest me for reporting to the United States, for using my English to proclaim that our tyrant was losing the battle for the dreams of Chile. So I spoke my final words into that telephone: "It's good to be home," I said, "and to be able to declare that not only the birds wake me in the morning, so good to be able to tell the world that my country is alive."

That was it. Done. Byline and out.

Except that a few seconds after I finished, a voice with a staunch Brooklyn accent piped up on the other side of the line and the hemisphere, the man in charge of the recording device in that remote Manhattan office: "Would you mind repeating that?"

And Ariel: "Repeat what?"

"The op-ed. From the beginning, please. Doesn't happen often, you know, but the tape got all messed up."

I had to reprise those words one more time, certain that Pinochet's agents would arrive before I got the chance to send them once again to New York. It was egocentric to suppose that the secret police were that interested in me, ridiculous to think they could understand immediately what I was saying in English and decide then and there to arrest me. But my inner voice in Spanish, the language in which they would come for me, was full of apprehension. What if that sound I'm hearing is someone's hot breath, *el aliento caliente de alguien*? What if the call has already been interrupted and I am speaking to an empty receiver? How can I tell if anyone is still recording this at the *Times*, while in a Santiago office a faceless man is barking out orders and a car is being boarded by men on their way to get me?

After the voice on the other end of the phone informed me that the tape recorder had worked fine this time, I hung up and sat there, still full of panic, trying to make sense of the lesson I had received, trying to frame this incident as a homecoming, a way of beginning to establish the frontiers of what is permissible, discovering how repression can shape the dusk of our every word. A lesson in the difference between exile and clandestinity, and in the slight distance that, under a tyrant, separates what we say to ourselves in the secluded fortress of our minds and what we say on the hazardous streets.

I had my English to thank for pushing me beyond what was prudent, for giving me a false sense of security, and ultimately for connecting me to the trepidation so many Chileans had been living with during my absence. So this is what people risk when they insist on flaunting their criticism and rebellion in spite of their fear, alone with their conscience and their skills, their cunning and their luck, unsheltered by the *New York Times* or the CBS cameras. My English was bizarrely telling me that no matter how much I wrote or published in that language, I would never be entirely sheltered, my body would always be exposed. English could not wrench me from Chile or the underprivileged world as long as I continued to speak out. It was not in hiding but on that slippery, unreliable surface of reality that power needed to be disputed; it was in the gray light of the day-to-day that the country had to be wrested away from the authoritarian government, that was where the immense majority of the people lived and where the armies of the night had to be defeated, and if voices like mine had to be heard, in Chile as they had been abroad, I might as well get used to being in danger.

And in the next few years, I somehow found ways to control my unbridled paranoia, to convince myself that fear could be conquered.

Typical of this delusion that fear held no sway over me was an experience in Santiago the day after I arrived there at the end of 1985, supposedly to settle in once and for all. I couldn't know that this intention would be thwarted six months later by the burning of Rodrigo Rojas, that I would have to wait until 1990 to again attempt a "definite" return. At any rate, back in 1985, maybe to prove I would not submit to the dictates of terror, I'd decided to join a group of women who were protesting in the Plaza de Armas. My friend Angela Bachelet — the mother of Michelle, the future president of Chile — had told me not to show up. "They've been very violent with men recently, whereas we . . . well, we always hope they'll treat us with more deference." But I didn't listen to her and was encouraged by how peaceful that circle of women looked, holding their hands up in the air. And nearby was none other than my own Rodrigo, living in Chile before starting his freshman year at Berkeley. There he was, wielding a camera for Teleanálisis, a semi-clandestine newsreel project.

I hurried to find my place in the circle and was greeted by grumbles—hadn't I been informed that this was a women-only event? I grabbed two uplifted hands, one on either side of me, muttering that I was an honorary woman, I was a feminist, had a female soul, who knows what other justifications. My psychiatrist friend Fanny Pollarolo was there, and Moy de Tohá, the widow of Allende's vice president, murdered by the military, and also among the group were a few relatives of the *Desaparecidos,* and they must have realized it made no sense to waste time trying to eject me when soon enough the *carabineros* would be ministering to the lot of us.

In effect, we had barely launched into the "Canción de la Alegría" when Rodrigo, who'd been panning over each protester with his camera, urgently warned me, *Ariel, vienen los pacos.* Through his lens he saw a mob of policemen in riot gear trotting in our direction. I hesitated—how could I run away, abandon these ladies? And then a woman to my left shouted, *Andate, huevón, que te van a sacar la cresta.* Get out of here, you idiot, they're going to beat the shit out of you. I released my hand from her grip and left the members of the fair sex to fend for themselves. From behind the safety of a tree in the plaza, I watched my *compañeras* being dragged off to police vans, some by the hair, some courteously escorted, some beaten, others allowed to go free—everything, as always, capricious and unpredictable.

We recounted the scene to Angélica a few hours later. Rodrigo had scarcely escaped himself, kept filming as he darted in and out of the melee. "But you, man, if they'd grabbed you, the one male in that group of women . . ." I responded by laughing it off, assuring him that I could always, with a bit of luck, land on my feet. I was glad that by lapsing into the absurd, dictatorial Chile had once again relieved the stress.

Supported by this false sense of invulnerability, I took ever-increasing risks during that long stay in Chile in 1986, as if I had vanquished the knot in my stomach when the voice in New York had told me that I would have to repeat my accusations against Pinochet all over again.

And then, on April 28, 1986, Halley's comet appeared over the Southern skies.

Approaching in all its magnificence, as close to Earth as it had been since 1910. A once-in-a-lifetime chance to see the cosmic visitor, and a once-in-a-dictatorship's lifetime to ask it to take Pinochet away and dispose of him in *el espacio sideral*, cast him into empty space, *llévatelo*, and never bring him back, *adiós General, no vuelvas nunca más*, see you seventy-six years from now. That gigantic burning body had already snatched Duvalier from Haiti and Marcos from the Philippines that very year, why not Chile next? Wasn't it the Year of the Tiger, when powerful men will fall?

I was one of some three hundred *manifestantes* in Plaza Italia, enjoying the joke while a phalanx of policemen kept at a discreet distance, confused as to how to react to something as apparently innocuous as a horde of citizens with fake telescopes waving farewell to the fiery nucleus and its dusty curved tail lighting up the Santiago night.

If only Rodrigo had been there by my side, but he was already enjoying his first year of college, so I had no one to tell me what to do, where to go, when two enormous trucks appeared, as if out of nowhere, and dozens of soldiers jumped out. They were led by a lieutenant, hawk-nosed, muscular, lean, his face entirely covered in black grease—one of the *carapintadas*, a fiend of the night. My group of protesters held their ground. "*Qué hacen acá, mierdas?* What are you doing here, shitheads?" the lieutenant shouted. "*Andando, andando, hijos de puta, conchasdesumadre*," a roar that was followed by soldiers advancing on us with a storm of kicks and rifle butts as we were herded up the street. "*Al trote, al trote, los hijos de puta.*"

I stumbled, sprawled, could feel a pain in my knee, my hand went to it and there was a gash in my pants. As I scrambled to get up, another kick left me reeling and then another blow to my torso but I kept hobbling up Avenida Providencia, away from Plaza Italia and the photographers and bystanders. I couldn't be sure if we were going to be arrested or shot or — but the glint-eyed lieutenant wanted to return with his blackened face and his pack to minister to other subversives. Was I safe? Outrun by my fellow protesters, I panted there, dazed and bleeding, until I saw that a lone soldier had been left behind.

Twenty years old, perhaps younger, obviously a recent recruit. He pointed his submachine gun at me, his finger trembling on the trigger. "*No se acerque,*" he barked. "Hands up, stay back, five meters, five meters away. *A cinco metros,*" he said. As I raised my hands, I understood that this young man couldn't really *see* me, my inept hands, my extreme defenselessness. Incredibly, *he* was terrified of *me*.

I saved my life by doing the one thing I'd been perfecting throughout exile: communicating with those who are different from me. Gently, as if this were the most normal exchange in the world, I asked him to look, I didn't have even a candy bar in my hands, I used the name of a sweet I had eaten as a child, *ni un Ambrosoli,* I said, a candy he must have chewed in his own childhood, as if that might remind him of his innocence and mine, and then I mentioned my two sons, one a bit older than he was, another younger — anything to establish a relationship, to come nearer than those five meters.

"Shut up," he said.

We stayed like that for a few more seconds, and I couldn't help it, my mind started dissecting, classifying, interpreting. Though I was five meters away from death I couldn't stop intellectualizing, during those seconds I allowed myself to become acutely cognizant of the divide between us: he was poor and uneducated and I was well-to-do and deft with Western words; he was of Indian ancestry and my folks noticeably came from somewhere else. Life had not been generous with him, and I wondered if, now that the tables had turned, he would make me pay for those centuries of neglect.

He didn't pull the trigger. He looked down at my bleeding leg, blinked, and I watched something brutal and sad drain out of his eyes. He breathed deeply as if ridding his lungs of a cloud and then veered his weapon away and made a motion with his free hand, I should get the hell out of there.

A few days later, on May 1, there was another protest, marking International Workers' Day. That morning, I'd put an extra pair of socks in my bag in case I got arrested. But then, instead of going downtown, I headed in the other direction, up into the hills with Angélica and Joaquín, and we sat by a river watching the turbulent water go by while Joaquín threw branches into the current and bombarded them with stones — and Angélica quietly held my hand. Gently as well, because it had been injured in the fall in the Plaza Italia. And there were two black-and-blue marks on my back, another near my ribs where a rifle butt had carved its message. And my left knee was crusted over with a scab, covering a cut so deep that I couldn't help but limp — even now, twenty-four years later, that knee still shows the mark. But the bruising of my ego was more painful.

What I felt on that May 1, 1986, was not the fear that hounded me when I was hunted down after the coup, and not the fear I felt when the man recording my voice at the faraway *New York Times* asked me to repeat my sacrilegious words, and not the fear of my own failings during that crisis of allegiance in Holland. What that lieutenant had hawked into me was unlike anything I had ever lived before, a trepidation encoded in my very body.

By 1990, when I returned to Chile, time and distance had somehow melted away the bone-deep fear. And I would have need of my newfound courage when I ventured into the land of secrets that Chile had become, when I dared to speak out in a land where most of the people were hiding from each other and from themselves, hiding from the truth they did not want to face.

Fragment from the Diary of My Return to Chile in 1990

SEPTEMBER 3

I am about to break my vow of silence.

Ever since I heard about the Comisión de Verdad y Reconciliación, while still in the United States, read about its goals, realized that the indisputable importance of its work was tempered by the agony it couldn't address, something began to stir inside me, a fever that has been hastened by our return to Chile six weeks ago.

And now I know what needs to be done: I am going to write a play.

A play that started out as a novel eight years ago in the United States and, as often occurs with my work, had its origins in an everyday, ordinary event, though its remote sources were percolating from way before.

During the first years of exile I had not desired—or needed—an automobile. Then, in 1977, we received an inheritance from Humberto, Angélica's father, who had died in London five years earlier. According to his wife, Lita, mother to Angélica's darling half-sister Nathalie, he would have liked us to have his old car. I drove it in Holland for a few months before it broke down on a snowy day and was abandoned forever to the scrap heap. But in the States, living in suburban Bethesda, a vehicle was of course indispensable.

I contacted Marcelo Montecino, a Chilean interpreter and prize-winning photographer living in exile. His cousin knew a mechanic named Pedro, from El Salvador, recently arrived in Fairfax, Virginia, who, besides selling me an inexpensive model, could repair it if anything happened to go awry. And, in effect, when the car Pedro sold me, a sturdy used Volkswagen squareback painted a discreet green, failed to start one morning, he swiftly came all the way from Virginia to fix the damn carburetor. During a conversation we had over coffee at our house, we were startled to learn that Pedro had been a sergeant in the Salvadoran army, which meant that he'd probably participated in who knows how many atrocities.

I liked Pedro, his calm reserve and unobtrusive sense of humor—and above all I liked having someone who could fix my car if I ran into trouble—so after he left I fantasized that he had abandoned his army post and his land in order to escape from those horrors. But on his next visit a few months later to patch up the brakes, it became clear where he stood in the conflict in his country. Not that he came right out and said it, he was as careful as I was to communicate through indirection and innuendo—he knew we were defenders of that other Salvador, Salvador Allende, and seemed to like us as much as we liked him. He had promised to keep the Volkswagen in shape, and he wasn't going to back down from that promise or our business, so both of us avoided mentioning anything that would transfer the respective conflagrations back home to our distant setting.

And what if our Good Samaritan were Chilean, what if, once I went back to my own divided land, my car broke down on a solitary road and someone just like Pedro happened by and saved my ass, what if I invited him back home and someone there, my wife maybe, discovered that this man had done, what if that man had . . . This mechanic, if he had been Chilean, would have been my mortal enemy, yet here I was partaking of cakes and ale (well, coffee) with him, someone who might well have murdered friends and relatives of my Salvadoran comrades Manlio Argueta and Claribel Alegría and . . . what if, what if the man who stopped on the road in Chile when my car broke down was not an uneducated sergeant but someone with the sophistication and sighs of a Jaime Guzmán as he listened to the Beethoven quartet in that theater in Chile, what if that man who stopped to help was hushing up harm done to a member of my family, to one of my friends, what if he had been the one who had approached Claudio Gimeno on the night La Moneda was bombed . . .

What if, what if, that's how my mind works, hundreds of bifurcations and characters and phrases jostling for attention, waiting for the instantiation when it all comes together, and I'm off, plunging into my next poem, story, article, play, novel, essay, whatever genre happens to feel right for the emotion dogging me, whatever is necessary to get it out of my system and onto the page, into the lives of readers, so I can be cleansed and they can be disturbed.

Sometime in the middle of the steaming summer of 1982, I started a novel tentatively called *Memoria,* uncoiling the idea that the wife of the stranded motorist had been raped and tortured by a Good Samaritan in love with heavenly music. The woman, recognizing the voice of her tormentor when he steps through her door, ties him up and seeks revenge. Outside that woman's house the thugs of Pinochet would be looking for their colleague. I imagined a Chile full of shadows and murk and menace, I imagined that woman's rage suddenly detonating, as mine had when that horrible Hübner man had dared to touch me in The Hague, as mine had almost burst out when Jaime Guzmán had sat down so close, so close, and sullied my Beethoven, my anger when I had read in that magazine in Paris that General Gustavo Leigh was laying claim to the music of my soul.

I wrote a first page of that 1982 novel, and a second one, and then tore them up and started again and threw those pages away, and continued like this for weeks, and something was wrong, something was most definitely wrong.

I was not ready.

First I had to return to Santiago, return once and for all and drink the bitter, hopeful river of Chile until the riverbed was empty.

It has taken me eight years to discover what was missing from that story, eight additional years of inferno that my country would have to go through before we reconquered our democracy, eight eternal years before I could sink into the ocean of suffering that is Chile.

I've spent these last seventeen years gathering all that anguish, a sponge of sorrows. I defined my role in this long voyage home as someone who dared to plunge into the abyss so I could come up with perhaps one fragile splinter of hope. That's how I built, word by word, my aesthetics of hope. During the dictatorship, all that ache had a dreadful communal meaning, and I assumed, perhaps we all did, that there would be some sort of catharsis when Pinochet was gone, a public reckoning. Well, he's gone now — or at least unable to continue repressing as he did a mere seven months ago — but instead of using that absence to deal with the dictatorship, the subject has almost disappeared. Now that the war is over, writers are supposed to go back to roses and birds and beauty. Turn the page, get on with life.

Except that I can't. I can't turn the page. Everywhere I go, it's enough to scratch the surface of each person, crack open the door to each person with a question or a look — that's enough for the howl of the past to inch out, a quiet howl, a hidden howl, a howl unasked for, neglected, a gasp more than a howl, but waiting for somebody to come and listen, somebody to care enough to listen.

It is out of that howl that Paulina has emerged.

That's what I'm calling her.

The story I interrupted eight years ago in the exile of Washington did not disappear from my landscape. It infected me and has, over the past few days, begun to take over my life.

Paulina burrowed into my heart, I see her alone by a roaring sea and frightened out of her wits because her husband is late, he is on the road somewhere nearby with a flat tire, waiting for a stranger to help him. Paulina is alone in that house, damaged as irreparably as the land that spawned her, yet fierce, so fierce that she won't leave me alone, or maybe I am the one who can't leave her alone, there is too much pain there for me to abandon that woman to silence, hers and mine, too many women like her, too much need for her voice to be heard.

Not, as I had thought, under the guise of a novel. The wall I was hitting my head against whenever I tried to resume the narrative has finally crumbled now that I have returned to the turbulence of this Chile without Pinochet's absolute power.

Paulina's deepest tragedy is not that she has been tortured. It is that she has been betrayed. The democracy she yearned for, where she could tell her story and receive justice, has instead slammed the door shut, demanded that she sacrifice herself for the common good. And this transition in Chile has also revealed to me the identity of the person making that demand: it is Paulina's husband, Gerardo, a lawyer on the *comisión* seeking truth and reconciliation, someone I could not have conjectured until now, during this perverse intermediary period, he is the one who, in order to find the missing bodies, in order to bring order and progress to Chile, must defend the doctor Paulina has captured in her living room. A conflict that needs to be enacted by real people on a stage — it is too critical to our sanity as a nation, too imperative and vital to languish between the covers of a novel that would take a year or more to publish. My words have to invade the physical space where the violations occurred, invade that space with the bodies that are being forbidden or concealed, invade the city where that doctor roams free, that city where a woman like Paulina must be asking herself if she will become like the monster who destroyed her life, a city where the lawyer, the hero and central protagonist of Chile's transition, tries to do the right thing for his wife and also for his country, a husband who wants to save his land and save the love of his life and who may end up losing his soul in the bargain. All three, Paulina and her husband and the doctor, will be onstage and also in the audience that watches, as the lies explode and the story of this fractured realm is stripped for all to see, for all to recognize. The story that everybody is living and nobody is talking about: this Chile where victims and torturers walk side by side, elbow each other on the bus or lick ice creams next to one another in the cafés, everywhere I go, everywhere I go, not knowing whether that taxi driver, or that heavyset man in the gray overcoat and the pinstriped tie, could be the one who approached Carmen Bueno on a street like this one, listened to Carmen's screams, sipped coffee and asked for more sugar and less cream and measured the electricity and the screams, is that the man?

It was necessary that I cohabitate and intersect with those people denied to me in exile, greet in this country someone like Jaime Guzmán, shake hands with someone like Hübner. The real reason I have come back: to tell stories against the flood of amnesia, force my hand to write about the most dramatic situation that history could have handed an author, the multiple scars nobody is interested in.

It is not a task I take upon myself willingly.

Wasn't this return supposed to be a time to learn from my fellow citizens, avoid my reputation for feistiness, refuse to stick a finger into every wound in Chile? And do I really want to get involved in the hectic swirl of rehearsing and producing and promoting a play instead of slowing down after more than twenty years of nonstop activity? Doesn't it make more sense to help the family settle in, now that we've just taken Joaquín out of Nido de Aguilas and need to find a different school for him, something more accommodating. Isn't our house here still falling apart, aren't family and friends awaiting multiple visits? And as for creativity, do I really want to continue on the road of immediate contingency, my work strapped to Pinochet, eternally attached to his fate? Haven't I been trying to escape his influence, declared my independence from Chile as the sole source of my inspiration in my latest fiction?

Am I to break my vow of silence just because the country is also silent? Surely there is somebody else in our land who is ready to tell this tale of treachery and retribution.

Or is there?

•

THERE WAS something missing in my life, in my life and in my story, a photograph, a very specific one, that I hoped to find when I journeyed to Santiago in 2006 to film the documentary.

Many years ago, in the worst moments of expatriation, someone had shown that photograph to me, and then I had lost it or squirreled it away so well that I could no longer find even a trace of its existence. None of the circumstances were clear, all was a blur in my mind except what stood out in the picture itself.

It had been snapped during the glorious times of Allende, and my face was a speck floating within its boundaries, one among a multitude of enthusiastic revolutionaries waving their fists in front of La Moneda Palace. I did not remember much more, not the date or the occasion or who accompanied me that sunny day, which of my friends and comrades. But that photograph was the only extant representational evidence that I had been there in Chile during the revolution, that I was not the ghost I feared exile had turned me into, and now that I was going to revisit my life for Peter Raymont's camera, I was determined to anchor that past in something just as visual, a piece of celluloid that could establish the distance between the Ariel of 2006 who looked at the photograph and the young man he had once been, that would punctuate the distance between the revolutionary who thinks he will never leave and the exile who suspects he will never go back.

The photograph, alas, continued to evade me during that whole 2006 trip. Friend after friend just shook their heads, no, it hadn't been them, though we pored over family albums and tore open old boxes in attics and closets: nothing, *nada*. Only Antonio Skármeta vaguely recollected a photograph of that sort, but he recalled it as pensive and forlorn rather than triumphant, and in any case he had no idea where it might be, if indeed he had ever possessed it. Ah well, maybe a positive spin could be put on the foiling of my quest, maybe the elusiveness of that photograph illuminated how the past can never be entirely grasped, certainly not in times of oppression.

And then a surprise: on the last day of the shoot, Peter Raymont, Rodrigo, and the rest of the crew ended up ferreting out the photograph from who knows where in Santiago, handing it to me in front of La Moneda, in the very plaza where it had presumably been taken. And my heart leapt for joy: there I was, there, right there, that long silhouette with glasses had to be me, though something inside warned me that nobody recognizable was nearby and the occasion did not correspond to my reminiscences and that man really did not look like me and . . . I kept on grinning anyway, posed with the photograph, pointed at the indistinct person I suspected was not me, I couldn't disappoint the camera, after all.

That scene with the fraudulent photograph is not in the film. Because two months later, through the good offices of the archive department of Julio Scherer's magazine, *Proceso,* the real photograph was found on the cover of a book published in Mexico in 1974, and that was really me, and nearby was Skármeta, and he had been right: we are indeed pensive, both of us, not a single fist in the air, no mouths rejoicing. I had, through the years, misremembered it, perhaps mislaid the photograph so I could misremember its lack of victorious symbolism, polish it clean of hints of defeat. Perhaps I didn't want it to contest a cardinal moment in that same plaza that I had not forgotten, even if not one photograph had been snapped of that event, even if the image of that night persisted only inside my mind and nowhere else.

I had last seen Salvador Allende alive one week before the coup, on September 4, 1973, when I had joined a million marchers who had poured into the streets of Santiago to celebrate the third anniversary of our electoral victory. That night it had taken our group—Manuel and Antonio, Cacho Rubio and Carlos Varas, and Jenny and Miguel Roth—it had taken us seven fervid hours to reach the street below the balcony of La Moneda where Allende was saluting the multitude. We had streamed by him, singing and chanting and unfurling flags, and brandishing sticks with which we swore to defend the revolution, and for one moment of eternal magic we convinced ourselves that we could still change the history of humanity, and then we looked back and saw him standing alone on that balcony waving a distant white handkerchief, and something unspoken and grievous made us immediately, without consulting one another, continue around the block and mingle with the next overflowing column, milkmen and women textile workers and ice cream vendors, so we could pass by one long, last time. Our hoarse voices might roar that *el pueblo unido jamás será vencido,* and that we would overcome, *Venceremos, venceremos,* but what we were really doing that night was bidding farewell to our president.

A week later he was dead, his body secretly dumped in a grave by the sea, the first of the *Desaparecidos* that Pinochet would hide away, the first of many denied a tombstone so they could be erased from memory.

Except that we had not complied.

More than thirty-three years later, on a bright December day in 2006, I fulfilled a dream I had been nursing during all those decades: I stood on that balcony where I had last seen Allende. The making of our documentary gave me access to that iconic space, the chance to stare out onto the empty Plaza de la Constitución, exactly from where our martyred president had saluted us, the same spot from which he had delivered so many speeches until the night he had waved goodbye with a white handkerchief. It was a poignant visitation of ghosts and memories, because down there now only strangers were walking; none of those who had shouted their defiance and hope that history would be on the side of the poor and forsaken were crowded into that plaza. Allende was dead and the revolution something pale and magnificent inside me, and it didn't matter if the photographs shone with victory or were twisted with defeat, it didn't matter if every possible photograph had been misplaced in some elapsed attic, because some solace drifted into me from that visit, all of Pinochet's repression had not stopped me from standing where Allende had stood, nobody has been able to quench these memories.

Or to stop me from believing in the justice of Allende's cause, his dream of a better world, the certainty that another world is indeed possible.

But how to arrive at the perfect society he envisioned — ah, that is another matter. The young man who had left Chile as an intrepid revolutionary, convinced that the end of capitalism was nigh and that any sacrifice was therefore justified on the road to socialism, was not the older man who stood on that balcony thirty-three years later.

When did it change? Was there a moment, a point of rupture, that separated one Ariel from the other, an event that marked the difference between these two versions, maybe avatars, of myself?

If I had to single out one day, it would be a glacial morning in the winter of 1982 when I was summarily expelled from the chancellery of the People's Republic of Poland in Washington, D.C. It had taken me many years to get to the point of walking through the gates of that embassy, to meet that ambassador under an ornate, bourgeois chandelier, it had taken me a whole lifetime to be able to tell that representative of Poland, undoubtedly a member

of the Polish United Workers' Party, that, as a socialist and fol-
lower of Salvador Allende, I was ashamed and outraged at what
his government was doing to the working class of his country in
the name of Karl Marx.

It had been the repression of Solidarność a few months earlier,
in December of 1981, the martial law declared by General Jaruzelski, the carnage in the Gdańsk shipyards, the outlawing of the
free trade unions and the jailing of thousands of supporters of the
Solidarity movement, the spectacle of the party supposedly embodying the hopes and desires of the proletariat turning its guns
on those very workers — those events had been *la gota que colmó el
vaso,* the straw that broke Ariel's ideological back.

My connection to communism had been, since the adolescent
start of my political education, an ambiguous one, probably because I was torn between my mother and my father and their con-
flicting visions of social change. Like so many of his generation
forged in the fight against fascism in the midst of the Great De-
pression and the debacle of capitalism, my dad had enthusiastically joined the Communist movement. Although, by the time of
my birth in 1942, he'd broken with the sclerotic and bureaucra-
tized Argentine party, he remained faithful to Marxismo-Lenin-
ismo, slavishly adhering to the Soviet Union and many Stalinist
practices. A loving, sensitive father and spouse, he was cold and
inflexible in defending measures that would bring about the presumed paradise of a classless society. Indeed, he considered those
birth pangs imperative, regretfully obligatory — a position he
maintained even after the Twentieth Party Congress denounced
the crimes of Uncle Joe Stalin, even after the invasions of Hungary and Czechoslovakia. For my father, outraged by inequality,
misery, and Nazism, those Bolshevik beliefs were the bedrock of
his immigrant identity; to abandon them would, I surmise, have
meant opening an abyss of introspection for which he was neither
ideologically nor psychologically prepared.

Whereas my gentle mother, a staunch opponent of the death

penalty and all other forms of brutality, had always been wary of communism's shortcomings. With a vivacious sense of humor that did not sit well with the commissars, she had founded the SRCLCP, the Slightly Reformed Conservation Life Communist Party, of which she was the sole member.

It was not as if there was much political discussion at the dinner table; my father would lay down the doctrine, the interpretation of current and past history, and my mother would merely offer some supple suggestion of her own that undercut the stiff views of the macho of the house without really confronting him. For my part, raised in a Chile with so many impoverished people, a Latin America scarred by arrogant Yankee interventions, I gravitated, indignant and flush with the Red self-righteousness of youth, to my father's positions, though always held in check by my mother's refusal to add one more victim to the century's long list of atrocities. So Salvador Allende was the perfect combination of my two progenitors: an admirer of Cuba and a fervent Marxist, he insisted at the same time that we could build a more just social order without having to repress our adversaries. Allende's democratic vocation fit well into my own personality and desire to avoid bloodshed, my fear of hurting any living being.

The failure of our peaceful Chilean revolution did not turn me into an advocate of armed struggle. On the contrary, it launched, inside me as well as in the left in Chile and across the globe, a multifaceted dialogue about how to achieve socialism in our time and not end up slaughtered by those whose power we were contesting. This debate would lead, in due course, to Eurocommunism, put forward primarily by the Italian and Spanish (and, to a lesser degree, the French) Communist parties, a sort of turn to social democratic ideals and strategies, a tendency I felt increasingly comfortable with.

At the same time, there were two realities of our Chilean struggle that could not be ignored. First, the international fight against the dictatorship was spearheaded by a Soviet Union pouring resources into the antifascist front we were trying to organize. And second, the Chilean Communist Party, because of its organizational skills, popularity, and experience of having outlived ferocious persecution in previous decades, constituted the backbone of the resistance to Pinochet. So even as I drifted away from the more rigid dogmas of Marxism, I bit my tongue whenever the Soviets or the Communists were attacked.

A balancing act that often led to ungainly pratfalls.

The most discomfiting of these occurred in Germany in the early spring of 1975. Our friend Freimuth Duwe, a left-wing member of the Bundestag for the ruling Social Democrats, and soon to be my editor at Rowohlt Verlag, lodged Angélica and me for two nights at his house in Hamburg and also arranged for us to visit Günter Grass's cottage near the city.

The Grass visit went smoothly for a while. The great author (and artist) showed us his latest etchings and then escorted us to the kitchen, where he was preparing a superb fish stew in our honor. Literature and Chile, Chile and food, food and wine and politics. I had devoured all his books, we were having the best of conversations, and he seemed more than willing to sign on to our committee to help artists in the homeland — everything chummy until he told me that he'd recently been in the south of France at a conference of solidarity with the Czech resistance to the Soviet occupation, which the Chilean Socialists had refused to attend. "Don't they realize, Ariel," Grass said, "that the Prague Spring and the Chilean revolution have both been crushed by similar forces, one by the Soviet Empire, the other by the Americans?"

I could glimpse my German host Freimuth trying to head me off, change the subject back to flounders and tin drums, but I wanted to treat Grass like a comrade: as a man of the left he'd understand that the Chilean Socialists couldn't publicly side with the Czechs against the Soviets and our own Communist allies. Günter's eyes narrowed and his bushy mustache bristled even more, if that were possible. What was my position on the Prague Spring?

I'd been in favor of that flowering of liberty and had condemned the Soviet invasion, as Allende had, to which our host replied that my current stance was then more shameful, because I was subjecting my freedom of opinion to petty party politics. Couldn't I see that Dubček and Allende were equivalent?

That's when things got rough. I didn't think we should equate the Soviet Union, for all its many faults, with the United States, and he shot back something about writers and their responsibility, and I was trying to be temperate, I was eyeing that luscious fish stew, but my mouth insisted that it was essential not to create a rift with the *compañeros* who were fighting in Chile, and Freimuth kept trying to run interference, but it was too late. I think that if Angélica had not been so charming and downright gorgeous, and if Grass's dog had not taken a liking to me, he would have thrown us out right that minute. Instead, he abruptly turned and began to work at his drawing table, on a figure he'd been engraving, and it was clear that our meeting was over, not a drop of stew to be savored, no more to be said.

Except when we timorously rose to say goodbye. He looked up and . . . "When something is morally correct," he said, "you must defend that position without concern for political or personal consequences."

What I could not tell Grass, would perhaps not have wanted to admit to myself, was that it was not only out of political pragmatism that I had turned a blind eye to the many glaring excesses of rulers who claimed to be inspired by Marxism. Beyond the loyalty and admiration I felt towards my father, I was held back by all those Communists by whose side I had fought in Chile and the rest of Latin America.

The dead, the dead, we carry so many ghosts to whom we often swear more fealty than to the living. The mythical martyrs of the Paris Commune who had been executed for trying to create a society without exploitation, the many heroes of my youth, as much a part of me as my cheekbones and my lungs, Marx in the British Library, the workers and soldiers storming the Winter Palace in St. Petersburg, the Vietnamese dying by the thousands to save their land and tell the world that a small country need not bow to bombs and bullying, *Ho, Ho, Ho Chi Minh, lucharemos hasta el fin.* I had chanted that in the streets of Santiago next to Pito Enríquez, thin as a toothpick, and he had died, my Communist friend Pito, in some aseptic hospital in Toronto, he had died of heartbreak, and he could not evolve, I could not tell him of my transformation and bring him along with me on this voyage away from the dogmas that he had believed in and that I had tolerated in the name of comradeship. And Enrique París who had been tortured to death, castrated at La Moneda, and Fernando Ortiz who had been picked up one overcast day in the Plaza Egaña and had never returned home, would never again smile at me as we crossed under the *álamos* of the Universidad de Chile. And the songs of the Spanish Civil War, *El Ejército del Ebro, rumba la rumba la rum bam bam,* the night when the Republican army had crossed the Ebro and defeated Franco's mercenaries, *ay Carmela, ay Carmela,* I had heard that song in the womb, in that same womb had listened from near and far to the partisans' *Bella ciao* in the hills of Tuscany as they fought Mussolini, and the Battle of Stalingrad and Fidel entering Havana, *ay Carmela, ay Carmela,* and inside me were Neruda and Brecht and Nazim Hikmet and Che Guevara, all dead, *rumba la rumba la rum bam bam,* so inflexible in their death and yet alive in the vast vocabulary of my "Internationale" heart, *arriba los pobres del mundo,* a part of the legacy I had inherited.

How could I simply throw out everything I had subscribed to and learned in my more ardent days just because of terrible things that were being done, mistakes and crimes and cruelty that I told myself did not, could not, nullify the quest for the equality that I was not willing to declare bankrupt? How could I renounce the foundation upon which to build a prophetic humanity, what I imagined could be our shining destiny, the epitome of fairness in that extraordinary phrase "From each according to his ability, to each according to his needs"? But above all I was held back by an acknowledgment of what our world would look like without the infinite struggle of so many ordinary men and women for racial equality and the rights of workers to organize, by what a dreadful planet this would be if those militants had not opened a space for women to be free, if they had not stood by the side of the unfortunate colonies of our earth on the road to liberation.

So it took a while, it took a persistent drip and drop and accumulation of errors and invasions and betrayals and mass murders for me to wrest myself away from that confusing allegiance, it took many misdeeds to get me, the personification of the fellow traveler, off the train. The Cultural Revolution in China had not been enough, and the Prague Spring had not been enough, and the killing fields of Cambodia had not been enough, and the appalling Soviet invasion of Afghanistan had not been enough, but every one of those events had chipped away at my armor, until finally, the workers of Poland—the WORKERS, *la clase obrera, los trabajadores, los pobres del mundo,* the wretched of the earth, no less—were being repressed for demanding the same freedoms being asked for by the people of Chile, and enough, *basta,* that was the tipping point.

Though exile as well as history had contributed to my transformation.

One morning in Paris—it must have been 1975—I knocked on the door of the apartment of Ugné Karvelis on the Rue de Savoie. Julio Cortázar, who lived with her, answered the door and told me they had a guest, just arrived from Prague, staying over for a few nights. Then Ugné slipped out of the living room and whispered to me that he was an author I would hear much about, a novelist destined for greatness.

It was Milan Kundera.

The saddest man I have ever seen.

That was no routine sadness etched deep into his face. It overwhelmed you with its aura, it extended out from him like a wave of unmitigated grief, a loss as inconsolable as my own but without any of the hope that I kept forcing myself to feel. He was soaking inside that sorrow as if it were his skin, as if he did not want to emerge from it; he wanted to use it, reach down into its depths, let it flow endlessly inside him.

He had fled from a regime that had become one of the cornerstones of the campaign against Pinochet, a regime I had refused to condemn publicly just a few months earlier at Günter Grass's house, a regime that was training our militants at that very moment, offering assistance in many forms and forums. But the arguments I had used in my discussion with the German writer would have been useless, harsh, futile, with Milan. Regardless of what you thought of the government installed by Soviet tanks in Prague (and I had no sympathy for its punitive dictatorship), the only possible reaction to a despair like Kundera's was a burst of compassion as cosmic as his own grief. The practicalities of "the enemy of my friend is my enemy" simply dissolved in the presence of a tragic disappointment that called out to me, that gave to a man persecuted by my allies a face, a real face, an anguish that I could identify with.

Less than a year later, at the International PEN conference in The Hague, when we expelled the Chilean center, I was nursing a mellow Armagnac at the bar of our hotel when I was approached by a tall, harrowed man sporting stubble on his cheeks, as if he had not shaved in several days, that contrasted with a carefully trimmed goatee. He was a Russian, and saved from Kundera's melancholy not by my sort of revolutionary enthusiasm but by a mischievous twinkle in his eyes, which had seen too much not to be amused by human folly. He was glad that Chile had been banished and hoped something similar could happen to the Moscow branch of PEN, though he doubted that the unanimity engendered by my nation would hold when it came to the Russians. He shrugged, as if used to this labyrinth of politics, and then unspooled to me how he had spent years in a psychiatric ward. "They said I had to be off my rocker if I was against communism, no rational person could be against communism, no rational person could ever write the verses that I wrote."

As he talked in his Slavic-inflected English, I wondered if, in fact, he was not a maniac, an insane asylum inmate posing as a poet; but no, he was a gentle soul and quite composed and measured. I had heard, of course, of the tactic being used by the Soviets against dissidents, of medicating and institutionalizing them. I had actually seen it as a positive sign, thinking that at least they weren't executing them, sending them off to gulags in Siberia. Not quite excusing such a deplorable pseudo-psychiatric practice but somehow accommodating to it—as long as I did not have to sit next to the injured party, drink something with him, with her.

As the years went by, I met many similar men and women. The Czech coach of a kids' soccer team in Amsterdam, a Vietnamese tailor, and poets from East Germany and Cuba who wanted only to be free to recite whatever words, obscure or magnificent, leaped from their mouths. As I steered through the crazy, fractious patchwork of expatriates and alliances, something in me began to alter: it became more difficult to pigeonhole each exile according to his or her affiliation alone. I recognized their stories, the wistfulness they felt for the way tea was brewed back home, all those humiliations so parallel to mine, their pride ground into dust just like mine. I did not need to identify with their political choices in order to absorb their pain.

My very slow opening to the victims of Communist experiments was accompanied by a parallel development: the progressive loosening of the bonds securing me to my own party.

Urged on by a combination of guilt and love, I had thrown myself during the first years of exile into an ever more frenzied engagement, the best antidote to the sort of dejection exuded by Milan Kundera, his solitary flight into the murk of exile with nothing but his mind in turmoil to accompany him. Nevertheless, as time sped by and I dashed from one incessant task to the next, a mad, intense pace of work that was sustainable only for those who were fiercely selfless, I realized with mounting relief that I was not made out to be a full-time revolutionary.

My shift towards a more voluntary, less addictive relationship with the organized Resistance became possible, I guess, because literature had rescued me from the remorse of silence. And inasmuch as the collective ceased to be the main buffer zone against loneliness, as the immediate traumas of the coup softened, a space began to open from which to establish a critique, not only of my party and all parties, but of the left in general. These were questions that had been quelled for far too long. Irreplaceable as these organizations might be in a war against a ferocious enemy with an army of its own, did they have to swamp every aspect of life, supply a definitive answer to every misgiving, force a choral answer to each and every problem? How to build a democratic society with parties that were self-perpetuating, as suffocating as the catacombs we were hiding in? Wasn't it time to abandon the Leninist hierarchical structure in order to lay the groundwork today for the sort of state we wanted to construct tomorrow — pluralistic, tolerant, humane, without a supposedly superior group of illuminati deciding the ultimate and totalizing truth?

And here was a question I dared not answer, let alone ask: Wasn't my own distancing a consequence of problems deep in the doctrine itself? I had deposited in a revolutionary party — true, of a new lefty variety — my liberty and conscience, and that Marxist philosophy, still a superb instrument with which to comprehend and critique capitalism, still a fiery vision of the future to which I continued to subscribe, seemed to have increasingly foggy responses to the new dilemmas of our times. What role do indigenous peoples play in guiding us out of our current crisis? How are we to integrate their ancient wisdom into the modern world? How to cope with the monster of industrialization — extolled traditionally by both the left and the right as the panacea to all our trou-

bles—without confronting the environmental degradation corroding our planet? And didn't it make sense to accept the world as
full of intractable mysteries, puzzles that could not be reduced to
one position? And the challenges of feminism and the sexual revolution and homosexuality and the new technologies and religion,
and the more questions I asked, the more I saw myself as a public intellectual at the service of all forms of liberation rather than
a militant subservient to the web and requirements of a collective,
the more I understood my political work as occurring mainly in
the uncertain territory of my writing and advocacy.

Even so, I believe that all the shifts in world history and all the
emotional and intellectual alterations in my life would still, given
my ideological baggage and the more intimate memories of victory and defeat, have been insufficient to make me ring the bell of
the Polish embassy with such public fanfare that blustery day in
Washington. One additional factor nudged me in the direction of
moral independence from the movement I had served with such
commitment. Oddly enough, this act of autonomy could only
have happened, for me at least, in the United States, the center of
the empire.

Because I was not alone that day at the embassy. That visit had
been organized by my friends Doug Ireland and Joanne Landy of
the recently formed Campaign for Peace and Democracy, a group
of American activists who opposed their own government's belli-
cosity abroad while at the same time establishing links with, and
offering support to, dissidents in the Soviet-bloc countries. Seven
years after rejecting Günter Grass's reasoning about the need to
concurrently denounce repression originating in both the United
States and the Communist countries, I had found a band of sisters and brothers that was impeccably dedicated to that very objective, a troupe (and not a troop) of radicals who understood that
one cannot be for freedom in Nicaragua and against it in Hungary,
that one could not deplore the U.S. support of Pinochet and applaud Jaruzelski's mentors in Moscow.

I had always been, in some sense, a human rights activist, but it was only through my reconnection to gringo rebels that I began to inch towards the man I am today. It was the American left that gave me space to think creatively, that offered a different sort of community from which to mature, a different sort of homecoming to the country I had loved as a child, that helped me find the freedom to walk into that embassy and shame that ambassador and feel the satisfaction of being ejected. If I had remained a member of the MAPU, I would have consulted with the hierarchy and toed the official line like a good soldier of the revolution. So that day in the Polish embassy I said goodbye to the soldier but not to the revolution, I bid a long goodbye to party politics but not to the politics of liberation, I stated that I would henceforth answer to my own conscience and no longer to the dictates or persuasions of patriarchs in any organization.

Other decisions would follow in the years to come. By declaring that what Polish socialists were doing did not represent me, I was preparing the way for other unsparing evaluations. I was able to celebrate the fall of the Berlin Wall and deplore the massacre at Tiananmen Square and condemn the execution by Fidel of Comandante Ochoa, one of the stalwarts of the Cuban revolution, the hero of the liberation of Angola, the man who helped defeat the apartheid South Africans. But it all started that day in Washington when I told the Polish ambassador that there can be no socialism without democracy.

Back home that night, I made a telephone call. Not to Günter Grass. Or to Milan Kundera. To my father in Buenos Aires. And told him about the day's events, how that Communist ambassador, who must have sung "Bella Ciao" and "Ay Carmela" in his youth, who must have been trying to be loyal to his own martyrs, his own dead stirring inside, I told my father how that Polish comrade of his had banished me from that building.

There was a pause on the other end of the line.

"Ariel," my dad said, and I couldn't tell if he was smiling or frowning. "You know that I disagree with you. And I'm sure you also know how proud I am that you're my son."

"I disagree with you too, Chebochy," I said, using the endearing term that was his pet name. "And you know that I'm also proud of you."

And we left it at that.

Ay Carmela, ay Carmela.

Fragment from the Diary of My Return to Chile in 1990

SEPTEMBER 5

Yesterday, we buried Salvador Allende.

Witnessing the calm, ardent crowds, the photographs that had been secreted inside mattresses all these years and held aloft as the coffin went by, an old man crying into a Chilean flag crumpled in his hands, the youngsters who had not been born when the military had given Allende a first, anonymous funeral close to the Pacific Ocean, those youngsters calling out his name as an incantation against death, the thousands upon thousands of the dirt poor that had lost a day's pay so they'd be able to tell their grandchildren that they had been here on this field of gentle battle, the elderly women in wheelchairs pushing themselves to Allende's new tomb as if it were heaven, the flowers falling from a millennium of hands. I wondered at the miracle of it all, how these *compañeros* had kept Allende's memory ablaze in the midst of their despair. I wondered at the millions of hours they must have waited for the day when they would be strong enough to resurrect the body we had kept alive in the forbidden darkness of our imagination, how they had been fierce and loyal enough to bring Salvador home.

And now real work lies ahead. While Allende lay in an un-marked tomb, we had to keep his myth intact and uncontested. A myth offers inspiration but also makes it impossible to hold a conversation with the man behind it. That is the dialogue that can begin now that Salvador Allende, so dead and so alive, has been returned to the earth we set aside for him all these years. Now we can start to live with him and without him, now his country can critique him and also seek ways to remain faithful to the vision of social justice and democracy for which he died.

And all of a sudden, lost and found in that forest of clenched fists and chanted slogans, I was visited by an illumination: in this passage from innocence to maturity that I have been obliged to navigate without Allende, I have not always shown the courage of my convictions. I had not been brave enough to die next to him in La Moneda and not brave enough to refuse my party's orders to leave Chile when the junta hunted me down. I had kept myself alive so I could tell the story of our era, that's how I had justified my existence each time I retreated from the utopian dream I still believed in, that's how and where I could redeem myself.

It's time to write that damn play.

I think I will call it *Death and the Maiden*.

·

WHEN WE had arrived in Baltimore in July of 1980, with our boxes of books and our dreams of migrating to Mexico, the portrait of Jimmy Carter greeted us in every government building, and it was almost inconceivable that he would not be reelected.

Not that I hadn't been primed for the triumph of Ronald Reagan less than five months after our arrival. Hadn't 1979, launched in the coincidental glory of the Sandinista revolution, the defeat of the Shah in Iran, and the earthshattering advent of our Joaquín, ended in global gloom, as Soviet tanks rolled into Afghanistan and Margaret Thatcher ascended to power in Great Britain? Her government, soon to be followed by Reagan's, was wretchedly reminiscent of the malignant Latin American laboratory I had escaped.

A sobering experience, to see the shock therapy executed in Chile being applied on the vaster scale of the United States: the same formulas of the Chicago boys and Milton Friedman and other ideologues who had been Pinochet's neoliberal gurus, the same dismantling of the welfare state and safety nets for the poor, the same tax policies favoring the rich, the same return to a ferocious Victorian Age laissez-faire market, the same smashing of the trade unions, the same visceral anticommunism and militaristic jargon. Although at least the American and British people were spared, as we Chileans were not, the accompanying horrors of dictatorship. But plenty of violence elsewhere: civil war in El Salvador and Angola and Mozambique, and soon the Contras butchering peasants in the land of Sandino, and slaughter in Eritrea, and the end of the socialist mirage in Portugal. And China, waking from the catastrophe of the Cultural Revolution and Mao's tyranny, was experimenting with totalitarian capitalism (Deng Xiaoping: "To be rich is glorious!"). As for Iran, it was persecuting trade unions and curtailing the rights of women. It had been the hostage crisis in Tehran that had undermined Jimmy Carter's presidency and incubated this American nightmare I was returning to, so many years after having fled from its first manifestation.

In effect, the fear that had haunted my childhood during the McCarthy witch-hunts seemed to be catching up with me all over again. With this difference: now I knew more about terror and how it is manufactured, I had seen what dread can do to a democratic land, and therefore recognized its manifestations in Reagan's America, was appalled that his policies could be carried out with the approval of an electoral majority. It was that Teflon popularity that depressed me most: it proved that this ultra-conservative plague was the dominant ethos of our time and that Allende's downfall, rather than the exception, had turned out to be the rule. And it seemed to whisper (or shout, perhaps) that the United States was, as I had trumpeted in my more radical days, beyond redemption or repair.

And yet it was the Gipper who, remarkably, kept baguettes on the table and the mortgage paid. Shipwrecked on the rocks of the Reagan counterrevolution, I had begun to transform what was a disaster for the people of the world into a bizarre source of revenue, publishing articles on the strange intersection of politics and culture in a country that had elected a former actor from Hollywood as its chief executive.

Those articles, and the books in English that started to accompany them, were not merely a desperate means of financial survival. The need to adapt to a new audience also forced me to re-examine my own assumptions and certainties, a process that continues to this day. It was one thing to write scathingly about the United States from the Chile of the revolution, and quite another to meditate and speak from the position of someone who had sought refuge in the very country among whose secrets I was rummaging. It is hard to be offensive when you live and shop and drive among the people you are intellectually dismembering, by no means the same to skewer amnesiac America's belief in its unique goodness and innocence when your own quest for purity has been sorely tested by a decade of revolution and repression, when you awaken each morning reciting the mantra of your own possible contamination. In Chile, the wild words employed in my writing were the ones in which I had chanted our collective future in the streets, the language I had once used to conspire with my comrades in smoke-filled rooms. Other men and women now occupied the place of those *compañeros,* and I had to ensure that my new foreign readers could follow me into the warren of my every interpretation. At the same time, I was aware of a zone of myself that I wanted to retain, a region of identity, no matter how fluctuating and brittle, that I was not disposed to abandon. I waged a series of skirmishes on behalf of that other recalcitrant Ariel inside, rebuffed the need to make the text so transparent and lucid that it lost a certain mystery and difficulty, and threatened to purge that vision of mine of the original swooping sin that had made it distinctive and interesting to begin with.

A struggle that, of course, cannot be separated from my adulterous love affair with English. It was in those initial years in the United States that I started to experiment with the language in which I write these pages, a language skewed and made slightly alien by the Spanish twin inside, marked by a rhythm and flow and twist, which is not quite how most Anglo authors habitually express themselves. I was joining others from around the world who were changing the language of Joyce and Salman Rushdie and Bob Dylan as they settled into it, made it theirs.

Not an easy voyage of discovery and addiction. I was venturing into the unknown territory of residing in both my tongues, the start of a process of rubbing one against the other, a labor of love that wanted to create one unified Ariel, one more linguistic mongrel on this planet. My gradual gestation of a new authorial self did not come without an inner fight over the pulse of each phrase, over an adjective supposedly misplaced or a verb that was meta-invented, over flow and vocabulary, style and grammar. Yes, my writing had to display how different I was, how secretly alien, yes, it had to stay connected to the source of its transgressive fire, yes, and my heart was going like mad and yes I said yes I will Yes — as long as this back-and-forth literary expedition recognized its limits and did not seek to undermine my ability to survive in the foremost nation in the world and the power its English afforded me.

A power that I used, once it had been unleashed in the early eighties, to full effect.

Back in 1965, Angélica and I had celebrated my twenty-third birthday, on May 6, by joining thousands of other Chileans outside the American embassy in Santiago in a protest against Lyndon Johnson's invasion of the Dominican Republic. While others cascaded stones and rotten eggs and vegetables at that building opposite the Parque Forestal, I launched—ever the pacifist—only verbal abuse, but if foul words could have killed . . .

Eighteen years later, on another May 6, a forty-one-year-old Ariel was again celebrating his birthday, this time by proffering to the holders of American dominion something other than insults: I spent that day walking the halls of Congress in Washington, handing out books that I trundled behind me in Joaquín's red wagon. Five hundred and thirty-five books, to be exact, one for each representative in the House, one for each senator.

The idea for that delirious political-literary project had come to me one insomniac night earlier that year of 1983 as I corrected the proofs for the English text of *Widows* that Pantheon, thanks to my loyal editor Tom Engelhardt, was bringing out. I can't deny that I was thrilled to be fulfilling my adolescent ambition to publish a novel in the language of my childhood, and doubly thrilled because I knew that having a book in the dominant lingua franca of our times might pry open the door to multiple translations and success in the literary world.

My glee was tempered, however, by the recognition of how far I remained from my original objective, conceived in Amsterdam, of using this very novel to sneak back into Chile, to reach my readers there. My old crone of a protagonist in *Widows,* that fierce woman who waited by the river for her disappeared men to come home, was about to make an appearance in the bookstores and book reviews of the United States and yet was still prohibited, as I was, from visiting her own country. Then it hit me, and the bouncy, ever optimistic, attention-deficit-disorder Ariel took over: Why not use the novel, newly minted in New York, to reach out to the legislators of this country? Maybe I could get some rich friends of the Institute for Policy Studies to buy the books at cost for my adventure, that's what I'd do. In my mind, I was already rehearsing the conversation:

"Good morning, sir, madam, Senator, Congresswoman, Congressman," and then explain that I couldn't go back to my land, but my plight was insignificant compared to what was happening to the Disappeared and to the relatives of the Disappeared in Chile, as portrayed in this novel I'm now dedicating to you, sir, madam, impossible to do something like this in my own country, because our Congress has been abolished and nobody has the right to vote there, just as I don't have the right to return. Perhaps, sir, madam, you could pressure the Chilean government into allowing me to go back to my country, perhaps you could protest the horrors of disappearance. And let me tell you about how the Reagan government bolsters the Pinochet regime.

And that was it, the spiel I repeated so many times during those eight days in May, rehearsing for what would be a flurry of interviews in the months and years ahead, my English hooked to the cause of Chile and Latin America and every oppressed nation in the world, my English articulated me into many meetings with minor Reagan functionaries and bankers, pundits and luminaries, and it was then that I started to cultivate an array of well-known stars, hitched them to the lone star of Chile.

It was a potentially perilous journey.

An inevitable distance began to yawn between the bright world whose rich and famous and powerful I was courting for the cause and the anonymous world of those being repressed in the darkness of places like Chile. My years of penurious wandering had joined me to those unknown people I was serving, but my anchorage in the United States and my assiduous cultivation of English opened up the possibilities of my participating, however marginally, however ambivalently, in the glamour of celebrity.

I got a kick out of this, that's the truth. It was a dream to meet Costa-Gavras in Washington when he premiered *Missing* and Richard Dreyfuss in Los Angeles and Arthur Miller in New York and Bill Styron in Connecticut. I lobbied Martin Sheen and Dan Rather, Jackson Browne and Jane Fonda and Harry Belafonte. I cultivated that craving of mine for vicarious exposure, while never allowing myself to lose sight of the fact that I was doing this work in order to save lives. Not a vague metaphor, as I was to learn when, in early November 1987, I got a call in my Durham home from María Elena Duvauchelle and Nissim Sharim. Along with seventy-five other Chilean theater people, they'd been warned to leave the country before the end of the month or be executed. The threats were signed by a squad calling itself Trizano — from a Chilean vigilante who had massacred Indians over a century ago — a blatant attempt to stop some of Chile's best-known figures from campaigning in the upcoming 1988 plebiscite. They needed a superstar to stand by their side on November 30, the day of reckoning.

I decided to write an op-ed for the *New York Times*. A few hours after it was published, the telephone rang. It was Margot Kidder. Alerted by Mrs. Solidarity herself, Rose Styron, Bill's wife, Kidder had contacted Christopher Reeve, her costar in the *Superman* films. "He's read your *Times* piece," she said, though I couldn't get it out of my head that it was Lois Lane speaking to me, "and he's fearless. I think he'll do it."

Later that day, the telephone rang again, and this time I recognized that deep baritone I had heard in movie theaters, the voice of the Man of Steel and of the evil dissembler of *Deathtrap* and of the benevolent lawyer in *The Bostonians,* the voice asking how he could assist the Chilean actors. At the end of our half-hour conversation, Reeve had a question: "How dangerous is Chile for someone like me?"

"Nothing would be worse for the dictatorship than for you to get hurt, but renegades from the secret police might decide to kill you in order to blame the opposition."

"And if I go, would this help my Chilean colleagues?"

"If you go, you'll probably save their lives."

After a pause, maybe four seconds long: "Then I'll go."

Reeve asked me to accompany him. I responded that my being there might cause him trouble, given my recent arrest at the

Santiago airport. And my presence would restrict the impact of his mission; it was crucial that he avoid a partisan stand. My wife, however, had volunteered to travel with him, an offer that was greeted by a long silence and then, Thanks, he'd be happy to have somebody guide him in a country that he knew next to nothing about.

It was an act of great bravery. Though many joked that Superman — a man who caught a child falling from a building, held up planes in the air, and stopped a dam from exploding — was flying to the rescue, the truth is that Chris Reeve's body was as vulnerable as that of any mortal, as he was to discover when, years later, he had an accident on his horse and was paralyzed for the rest of his life.

But Chris confronted that desperate tragedy in the same way he confronted the challenge of Chile. If fate had granted him so much renown, it would be a sin not to use it wisely. He could save real lives — not in dark movie halls, but in the darker territory of history. And he did not flinch when, the day after he arrived in Santiago, the government prohibited the planned public act of solidarity with the endangered actors and the site of the protest was moved to a claustrophobic warehouse, the Garage Matucana. Undeterred, he entered that space where thousands had hazardously congregated. I followed these events breathlessly from afar, informed by Angélica, who was at risk herself, as courageous as Chris, my Angélica, as courageous as those actors, all of them joining together in an act of monumental defiance.

Not his last service to Chile.

The campaign for the No vote against Pinochet in the 1988 plebiscite was gearing up, and I enlisted him to appear in one of our publicity spots that the dictatorship had been forced to offer us every night at an ungodly hour, but that all of Chile had tuned in to, and there was Chris again, telling the TV viewers not to be afraid, that they were not alone.

Nor was he alone.

My mission for the plebiscite was to bring in as many celebrities as I could find. And I delivered — *they* delivered — and I was thrilled to be *valuable,* included in the campaign, and tried to suffocate the unsettling thought that I had spent so many decades denouncing the way fandom emasculates citizenship and postpones responsibility, the way Hollywood manipulates our emotions and tamps down our rebellion, I had spent a good part of my life calling for ordinary humans to emerge from the shadows and into the light, and yet when the chance had come, I relished being in touch with the luminaries, caught in the web of their enchantment. Yes, it was for a good cause, and yes, I was giving them a chance to use their fame for something meaningful, but the elation I felt was . . . well, unseemly. When I contacted those stars personally, voice to voice, face to face, I could sense my ego swelling, puffing up in a puerile delight that crept out of some deep-seated insecurity intensified by years of rejection and exile.

I was to be taught a lesson in humility one crazy midnight in October of 1988, shortly after the people of Chile resoundingly defeated Pinochet in the plebiscite. "He has an eye that can see everything," an old woman confided in a whisper when I went canvassing for votes in a *población* a few days before the referendum. "He'll follow me into the booth, take away my roof, but I'll vote against the man anyway. This is my one chance."

And what better way to celebrate that she and so many millions like her had chosen the path of democracy, what better place to rejoice than in a stadium, listening to the blur and roar and wave of seventy thousand screaming, heaving rock fans at an Amnesty International concert, one week after our triumph at the polls? True, it wasn't Chile yet, we'd need another two years before we could hold that concert in Santiago. But Sting and the other singers on that World Tour for Freedom had added that show at the Estadio Mundialista in Mendoza, Argentina, as a way of calling attention to the struggle in Chile, where of course Pinochet would not have allowed even one of the troublemaking musicians to disembark.

Mendoza, just on the other side of the Andes, was the city from where the ragtag army of San Martín set forth in 1817 to liberate Chile, so why not create a new Ejército Libertador, thousands of Chileans crossing the mountains to Argentina to celebrate human rights?

Sting had asked that before I introduced him to the multitude, I declaim one of my poems — and I willingly agreed, feeling that I somehow *deserved* that honor. I had helped to organize this wild extravaganza, Sting's "They Dance Alone" had been partly inspired by my poetry and fiction, this seemed like a unique chance to cast my forbidden verses into the ears of all those young Chileans.

The show's producer, Bill Graham, was dubious. "Those kids will have been out in the sun for hours," he said, "and then in the cold of night waiting for Sting," they would probably not be in the mood for poetry. But Graham asked for the poem anyway, and nodded after reading it, and he said yes, if I was willing to risk it, poetry's always good.

I entered the stage under a flood of lights, enough to make me sweat in the chilly Mendoza night, but I sweat even more when I realized that the spectators were expecting Sting, not this unknown Dorfman fellow. The dazed puzzle of silence was speckled with some hoots, and I hesitated. Wouldn't it be better to ditch the whole poem thing and just announce, like a crazed incarnation of Jack Nicholson in *The Shining*, here's . . . *Stiiiing!* I looked at the real Sting, standing just out of sight of the crowd next to Peter Gabriel, and they gave me a reassuring smile, they'd tamed this sort of monster many a time, and I ludicrously remembered the scene in *The Way We Were*, one of Angélica's favorite films, when Robert Redford says to Barbra Streisand, "Go get 'em, Katie," and some voice inside me, not Redford's for sure, was saying "Go get 'em, Ariel."

I took a deep breath and went and got them, or tried to get them, because they seemed to be the ones out to get *me*. As I launched into the first verses something stirred in the loins of the stadium, and it soon became an uproar. I could hear—or at least hoped I was hearing—shouts of encouragement from the sons and daughters of friends out there, Queno's son Matías, and Rodrigo and his mates, and I kept plodding on, the boos and protests now so loud that I couldn't hear my own words, my litany against death and disappearance trying to soar above the clamor from without and the self-questioning from within, what right do I have to come between these fans and their idol? Why did I ever want to be surrogate to a rock star and bathe in his aura, how could I have become so addicted to the fawning of the world and the whirring of the cameras, what insane delusion gripped me, why, why? Even so, I continued. Pinochet couldn't silence me, and this crowd won't silence me either, and then I'm on the home stretch, almost done, I've finished, I can finally gesture to one side and yell with my last lungful of air, *Y ahora, Sting!* And he came out with his band and hugged me and I staggered out from below the diabolical lights and Bill Graham gave me a friendly punch in the arm. "That was great, man, that was really gutsy."

I was hoarse and chagrined, but there were more pleasurable matters to attend to, the rollercoaster ride of the music itself had started, and hey, I'd never been to a Sting concert, so I circled around, with my all-access backstage pass hanging from my neck, to join friends and family in the section right in front of the stage and Angélica offered me a comforting kiss and Verónica de Negri a hug. The mother of Rodrigo Rojas, burned by the Chilean military, had been on this tour for the past month and a half, demanding justice for her martyred son, and she'd soon be up there onstage with the women of the *Desaparecidos,* dancing with Sting and Peter Gabriel instead of dancing with the dead and the missing and the invisible ones.

Finally that was all that mattered, this celebration of our victory in the plebiscite, the fleeting solace of the songs. I was able to leave my public disgrace behind and lose myself in "Roxanne," *you don't have to put on the red light . . . don't have to sell your body to the night,* and then there is the finale with everybody up there singing together, Sting and Peter and Tracy Chapman and Youssou N'Dour and the Inti-Illimani, *get up, stand up, stand up for your rights,* Bob Marley alive in Mendoza telling us, all seventy thousand of us, that heaven is not under the earth, urging the world to remember that the story hasn't been told, *get up, stand up, stand up for your rights.*

It took me a while to probe — with the help of Angélica, who doesn't care about fame and cares less about famous people — the meaning of that public pillorying, to realize that a slap in the face can become a good slap if you can only come to love it, embrace its pain as a learning experience.

A lesson that was to serve me well when the extraordinary success of *Death and the Maiden* opened even more doors, put me in touch with innumerable actors and directors and musicians, not just to enlist them in human rights campaigns, but also as collaborators in my artistic work, my plays, my films. A mantra I would need as I was given access that would have made the poet who introduced Sting that night in Mendoza dizzy with envy, a mantra I try never to forget: This is not about you, Ariel. Treat the exaltation of the limelight as ephemeral, approach every figment of fame with modesty and, if possible, with self-effacement.

A refrain I would repeat to myself when *Death and the Maiden* triumphed on Broadway and was made into a film by Polanski, and I repeated it when Bono dedicated a U2 concert to me as he shouted out my name. I recalled my ultimate insignificance when words I had written were pronounced onstage by Meryl Streep and Sean Penn, Kevin Kline and Julianne Moore, Sigourney Weaver and Alec Baldwin, whispered to myself: This is not happening because you deserve it; you personally don't deserve anything. This is not, it should not be, about you.

Do I feel proud? Do I continue to crave attention? Of course

I do, of course I pinch myself when I spend weeks with Viggo Mortensen adjusting phrases and nuances while we rehearse a new play, or when I rise to address the UN General Assembly and read the delegates a poem, or when I deliver the Mandela Lecture in South Africa, and I say how can this be happening to the writer who had to tap out his despair on a toilet seat in Paris and was expelled from a first-class compartment on a train to St. Denis and spent decades begging for help, this can't be happening to the refugee stranded without a home or a visa or a job in a hostile country, this success can't be real, these photographs that capture me with presidents and Nobel laureates and superstars. But as soon as I perceive a hint of cockiness and entitlement creep into my soul, I quickly cut myself down to size with the message of Damocles from the Mendoza debacle, remember that you are mortal, remember that this is not, it cannot be, should not be, about you. I tried not to forget that refrain when Peter Raymont's documentary on my life was shortlisted for the Oscars or I won the Olivier Award, this is not about me, I will be gone soon and all these wonderful actors and directors will be gone soon and what will be left, if anything at all, oh vanity of vanities, is what we created together, what may be left, if we are lucky, is some trace of faint beauty, some slight hope of kindness prevailing, some person somewhere who has tasted freedom and has been inspired to yearn for more freedom, more beauty, more justice, I am only a vehicle for a force that is in the universe and is blowing through me. It's all right, perhaps inexorable given my personality and my damaged past of exclusion, all right to flit in and out of that world of celebrities as long as you don't believe in the flashing bulbs or the hype or the praise, as long as you remember the man who learned a lesson about loyalty and truth in Holland, the man who learned a lesson about terror and truth from the anonymous people of Chile, as long as I never forget the lesson about narcissism and truth that man learned in Mendoza on a night of humiliation that I now recall as a blessing.

Fragment from the Diary of My Return to Chile in 1990

OCTOBER 8

This morning, I completed the play.

It's been night and noon and every hour in between, squeezing any extra minutes out of the day in order to have it ready for staging before I leave for Duke three months from now, I've never felt so possessed by a story.

I've ended up writing a thriller.

In that isolated house by the sea, in an atmosphere as claustrophobic as any Agatha Christie setting, Paulina subjects the man she believes to be her tormentor to all manner of indignities, insists that all she wants is the truth from him. Dr. Miranda protests his innocence and is defended by Paulina's husband, Gerardo, for a variety of reasons. What would happen to the country, to the rule of law, if aggrieved citizens engaged in vigilante justice? And what would happen to the transition, to its delicate balance of power that could be so easily upset, to the pact between former enemies that has guaranteed impunity to the followers and accomplices and executioners of the dictatorship? And what if this Roberto Miranda is not guilty, what if Paulina has misidentified the perpetrator, is attacking the wrong person with circumstantial evidence that would not stand up in a courtroom? What if her thirst for revenge ends up destroying the Truth and Reconciliation Commission and its ability to heal the wounds of the land? But behind the legal, philosophical, and political arguments that Paulina's lawyer husband wields are his more personal motives: How will this "trial" affect Gerardo's career, compromise his meteoric rise and possible appointment as minister of justice?

And what about Paulina?

Paulina must decide, during one long day and night, who she really is, if her pain has turned her into somebody like the doctor who played Schubert while he raped her, or if she is able to step outside the cycle of violence she did not initiate but in which she is nevertheless trapped.

Of course, before she can even ask herself the fundamental question about her identity, she needs to outwit both the doctor and her husband, she needs to set a trap for them so she can reclaim her Schubert as a balm to the better part of her soul rather than a reminder of the perversity that men are capable of in the worst circumstances. Only then can Paulina's ordeal, and the play itself, come to an end, only then can those in the audience ask themselves those very questions about their pain, their complicity, their country.

The audience? While writing the play, I had occasion to summarize the plot for some friends, a couple of them quite well placed in the corridors of power. After the hackneyed smiles of encouragement, a sort of hesitation crept into their eyes and curled out of their lips and . . . what they said and did not say, merely implied, was that the story I had described might turn out to be, well, inconvenient. Maybe the country's not ready, maybe you should wait, what if it isn't the right time for this sort of provocation?

Not the right time?

When I confided these misgivings to my psychologist friend Elizabeth Lira, she responded that the people now in power fear me. Me? Feared? Yes, because I know them intimately, the elite of this country, the men who led the Resistance, I'm projecting the doubts they themselves harbor about this transition, reminding them of the price they have to pay for stability, the qualms they cannot afford to recognize.

So Chile has no place for Paulina. But she isn't dead. She may be deserted but she isn't dead—and I, for one, with my words am going to bring her out of the darkness.

Not just you, Ariel. Because what better sign that you are back from exile than to have written a play that requires a community to stage it and a community to receive it and a community to support it, a play written and spoken in Spanish and simmered in the same language of rage and hope that is being suffocated in the streets of our city, what better homecoming gift?

And yet I cannot ignore the world beyond the borders of this land either.

My labors are not yet over.

I still have to translate the play into English.

That's the pact my languages have subscribed to, the truce they reached in their war for my throat. These last ten years in the United States have brought me to accept the glorious plurality of my bilingual existence. It may have been in Spanish that *La Muerte y la Doncella* has been conceived, in Spanish that Paulina was tortured and Miranda swears he's innocent and Gerardo looks for a way out, but English has not been absent from the creative process.

Indeed, perhaps it was only because English was inside me, exile was inside me, whispering hints as the play was being written in Spanish, perhaps because distance has attended this birth, even if I am geographically installed here in our house in La Reina — maybe that is what has helped me to avoid the traps of realism. My play has not, after all, depicted what happened in Chile but in a country of the imagination where Paulina can have some measure of justice, at least in the moonscape of her mind, and this makes the play, perhaps, meaningful to other places around the world that, after the fall of the Berlin Wall, are going through parallel processes of transition.

At the end of November, the Institute of Contemporary Arts in London is holding a week against censorship during which, one evening, they'll stage my play *Reader*. But *Death and the Maiden*, as it will be called in English, would seem to be more appropriate, maybe they'll perform that, recognizing its relevance. Which would allow me to hedge my bets, let my work spread its wings in the wider world where I have spent most of the last two decades. Just in case Paulina's story really is inconvenient for Chile. Because I can't help but wonder if this expedition into the madness of this country only has a future outside the country itself, what if the success I thirst for will be in English and not in Spanish?

What if nobody wants to stage it in Santiago, what if they are too scared?

•

BEHIND THAT rush to showcase the play abroad, to arrange a reading in London, there were problems I didn't want to admit, a conflict with the cultural elite of the country that had been smoldering under the suface of all those visits before our 1990 return. I had prefered to ignore a residue of resentment, a hint dropped now and then regarding the luxurious lives exiles led, snide comments about my *éxito*, how unfair, some people kept saying, that so many deserving Chilean authors were unknown abroad. I had brushed these comments aside, chose not to ask myself why such scant appreciation was expressed by those who had most benefited from my solidarity.

I had not done my work in order to be thanked, at least that's what I believed in exile when I'd answer a needy telephone call at dawn, when I sent funds to artists back home or organized trips for them abroad in search of support, when I opened doors and wrote articles and lobbied the powerful. My model had been Dersu Uzala, the hunter in Kurosawa's film who leaves food and water and blankets in a Siberian cave without ever knowing who will be fed and warmed in the future because of his charity — that's how I projected my selflessness. But the moral of the fable comes a lifetime later, when the protagonist, feeble and old and lost in a blizzard on another part of the tundra, stumbles upon a refuge where some nameless person has left succor for distressed travelers. Just as Dersu had in the past. So I guess that, after all, I was expecting some measure of reciprocity, I nursed the subconscious notion while in exile that, back in Chile, I'd be greeted with open arms.

Instead, most of the people who had any cultural or political power, including many of my former comrades in exile, turned their backs on outsiders like me. They had set their sights on ascending in status, becoming insiders, they were clawing their way into society—a desire that would be symbolized, as I saw it, by their inclusion in the sanctified halls of *El Mercurio*.

El Mercurio has been, since its founding in 1827, the conservative voice of a conservative country, distilling its presumed wisdom with dry superciliousness, the ideological mortar holding the bricks of traditional Chile together, the arbiter of our land's destiny. Rabidly anti-Allendista, it had used CIA money provided by Nixon and Kissinger to spearhead the offensive against our president; it had been more responsible than any other entity in Chile for bringing him down. Then, during the dictatorship, its owners were among the staunchest supporters of Pinochet, deriding the relatives of the *Desaparecidos* as liars, imparting from arch editorial pages the reactionary direction in which the regime needed to go. It was the opposite of everything I stood for, yet it was indispensable to read because of its importance and, in effect, its journalistic excellence. *El Mercurio* could ignore you, but you couldn't ignore *El Mercurio*.

Even abroad I couldn't escape its allure, and as I read through the pages my mother-in-law sent, I remarked upon an intriguing shift over time. *El Mercurio* had always flaunted a society section, a few pages where Chile's *dinerati* paid to display baptisms and weddings and corporate milestones. As the prosperity of that opulent class grew with each Pinochet decree privatizing the wealth of Chile, those pages — called *Páginas Sociales* — also grew disproportionately: dozens of group photographs of cocktail parties, openings of clinics and anniversaries of boutiques, a Who's Who of

Chile's *haute monde,* only the right sort of people, only supporters of the military government.

And then, from the plebiscite of 1988 onward, drip by drip, face by face, the democratic opposition that was on its way to power started to filter into those windows of social life, this figure and that one and this other woman, discreet appearances at first by those people soon to be in charge of the country but who had gone unmentioned in the official newspaper for decades. Their gradual presence in high society suddenly became gossip-worthy, this guy, that guy, this wife, that lover, this executive next to this former exile, this colonel sipping a martini close to that other former political prisoner, Hey, I saw that you were at the inauguration of Suzie's art gallery, *claro que sí,* and I saw you were at that alumni dinner, I didn't know you had graduated from St. Margaret's British School for Girls. The former pariahs lining up to have their pictures snapped and their lives validated by the right-wing media, given a certificate of good manners and *buena conducta* by the owners of Chile. Such breathless zeal to be ushered into the enemy's jet-set pages, that stampede towards half-baked eminence by what Angélica sardonically called the Red Set, that herd of dissidents clamoring for the limelight, was disconcerting, or maybe nauseating is a better term, given that the price of admission was to refrain from comments that might discomfit the hosts.

These *páginas sociales* and what was christened *la taquilla,* the in-crowd, this *danza versallesca,* came to epitomize for me the new Chile that was rising from the ashes of the dictatorship. It was one where friends and foes could coexist for the good of the *patria,* an uneasy alliance between those who wanted to forget the past because it was full of their crimes and those who wanted to forget it because it was too painful, because remembering that terror too insistently could lead to its repetition, and so the word *dictadura* disappeared from our vocabulary and the more neutral *régimen* took its place, and it became vulgar to refer to *los pobres* or *víctimas,* as if not speaking about something could cause it to vanish, the winners and the losers (but who won, who lost?) mingling under the hushed chandeliers, no atrocities under the soft glinting lights, not when the tea is being served, a country where it is obscene to mention sexual organs on TV but not obscene to have applied electricity to them in some nearby dungeon, a country where abortion is outlawed but not those who aborted a country, *ay qué mal gusto,* where are your manners?

Like my protagonist Paulina, I wanted no part of that world, that compromise, that room where women come and go talking of Michelangelo. Nevertheless, in some sick zone inside me there crouched the Ariel who had stood onstage in Mendoza with Sting and who now wanted to be recognized and feted by *El Mercurio,* the Ariel who, after so many homes abandoned and friends left behind, only wanted to belong, an Ariel who somehow assumed he was a member of this club and entitled to talk of Michelangelo. Except that I knew, and those who did not invite me must have known, that at some point I might mention torture, rats in a vagina, and crushed genitalia, *ay qué mal gusto,* how tasteless. No wonder the door was being slammed in my face. I was a nobody, a *don nadie* easy to ignore with impunity, and I couldn't have cared less and I couldn't have cared more, I craved to be included and was delighted to be excluded. *Apareciste en El Mercurio?* Did you appear in *El Mercurio?* No. I *disappeared* from *El Mercurio.* In a far more benign way than so many of my unfortunate compatriots, I had also been *desaparecido.*

And now, twenty years later, I can appreciate how lucky I was to be airbrushed from the ballrooms of the new elite. That omission nourished *La Muerte y la Doncella,* forced me to stay connected to Paulina's fiery voice. If I had not been cast from the halls of power, it would have been more difficult to hear Paulina's outrage and the seething, hidden story of Chile.

What a favor they did for me, those who forgot that I existed, forgot me so I could remember who I really was. It was an estrangement that Paulina would reveal to me as she grew inside, what that fictional being whispered to me from a pain that was real: there's no place in Chile for someone who speaks his mind when most people want to shroud their emotions, no room for a writer who has spent far too many years contesting silence to yield to its lies now. Now more than ever, when nobody can jail or banish you or kick you out of a job, now when the only thing that can stop you is yourself, your wish to be loved, sneaked into a photograph, paraded with the glitterati; now more than ever democracy needs to be strengthened by dissidence and criticism and ambiguity. I'm a man who would rather say I'm sorry after I've done something than ask permission before I act. I probably would have written the play anyway, anyhow, even if I'd been hailed by the official world of politics and culture, given a role or at least accorded respect by the new government, incorporated into rituals and gatherings and photo ops of the crème de la crème presiding over Chile. But fortunately they never gave me the chance to be tempted.

Why was I hurt by their rejection? Why did I keep banging on the doors of people who had been my friends in exile and before exile and now had power, like Enrique Correa, who was Aylwin's éminence grise, the president's right-hand man? Didn't I realize that by leaving the party I might have gained independence but had lost clout, access, a brotherhood that would take care of my interests? Why did I suppose that in the Chile where Allende was safely dead there would be a welcome mat out for loose cannons and utopians like me? Why did it take me so long to realize that unpleasant truth?

I had been spoiled.

When I went back to Chile in 1990 there was a memory inside me, and it's inside me now that I know I will never go back to live there, it's here in me, that welcome-home dinner hosted by a hundred friends and *compañeros* on our first return from exile, in September of 1983.

In better times Angélica and I had lunched at La Querencia, a sprawling restaurant nestled in a ravine way up in the hills of Santiago. A place of such quiet in the afternoon that one could hear below, half hidden from view by a copse of lovely weeping willows, a brook with water so clear that it seemed primeval in its purity. At night, though, it was the noisiest spot in Santiago, where clusters of celebrants guzzled mediocre food and danced to an orchestra of blaring horns and outrageous percussionists.

Our ten years of absence had not changed La Querencia. The same orchestra seemed to be performing the same boleros and rumbas for the same hordes of boisterous merrymakers in the large main dining room. Our party had reserved an annex with movable frosted-glass doors that kept prying eyes out, but not the unbearable racket.

We greeted our fellow guests of honor, José Antonio Viera-Gallo and his wife, Te, who had just returned from their Italian exile. In the midst of our hugs, he yelled out — above the din of the music — something about a rowboat.

"What?"

"The rowboat finally brought us home, just as you said it would!"

The last time Angélica and I had visited José Antonio and Te in Rome, back in 1978, ten or twelve of us had ended up, after a raucous dinner, on the wooden floor of the Viera-Gallos' small apartment, where I had insisted that we all start rowing, that if we put enough muscle into it we would arrive in Chile. So the company took up invisible oars and heaved to, chanting our hope into the imaginary wind, crossing the Mediterranean and past the Pillars of Hercules, look, there's Gibraltar, and then into the Atlantic and the Cabo de Hornos and the Straits of Magellan and up Patagonia to Valparaíso, and we kept on rowing on land, like resplendent galley slaves, rowing for our lives, rowing ourselves out of banishment, willing ourselves home. And now we had made it, we had joined our friends, not in any of those remote, alien cities, but here in La Querencia — a word derived from *querer,* to want and also to love, *querencia,* which means longing and is close to *cariño,* affection, this is where we are loved.

I hardly noticed, as the evening progressed, what we were eating, what we were even saying — as long as I could feast my eyes on each face, all those who had struggled for our return, to keep some semblance of a country safe for us. And then Jorge Molina, a lawyer prominent in the defense of human rights, tinkled his half-empty wine glass with a spoon and stood up, informing us wryly that he'd be brief, because the band had awarded him merely a three-minute reprieve to deliver his speech.

He told us how much those who had stayed had needed us all these years. We have grown, he said, and you have grown, and we know that what you have lived out there, in that world, has made you different from us, but we have to believe that not that much has changed. We may not be the same people of ten years ago, too much has happened, we've seen too much, but still you can recognize us. And now, he continued, we've earned the chance, we've fought for this chance, we've paid for this chance, *esta oportunidad,* to begin a process of integration, for us and for you, it's time to bring together our sameness and our differences.

The simplicity and truth of his words were enhanced by the calm lull of intimacy, the first time that night that our ears weren't being bombarded with the blaring music. As if on cue, as if some hidden maestro had been eavesdropping, a cha-cha from the next room exploded. A roar of approval! The rest of La Querencia's clientele had not appreciated being cut off from their own version of communal joy. A thumping of feet next door indicated their intention to make up for the lost three minutes with an exacerbation of their previous bacchanalian efforts. Abruptly, the glass doors swung open and a snake dance of young women burst into our room.

"*La novia,*" they shouted. "*Queremos a la novia!*"

The owner, afraid of the repercussions of hosting our troupe of flaming dissidents, must have informed the other guests that we were holding a wedding party, much to the delight of the bache-lorette amazons bidding farewell to one of their brood before she tied the knot. They selected Te as our lucky bride. Wisely, she ac-cepted her assigned role and signaled for the rest of us to join in the fun.

Liberated onto the dance floor of the main dining room, our small crowd let loose all the accumulated tension of our dinner, dancing, not in tow or coupled up, but in a manic, delirious jum-ble, until the band launched into one of the all-time favorites, "Ali Baba y los Cuarenta Ladrones," to which we sang along at the top of our lungs, repeating the main stanza: "*Ábrete, ábrete, Sésamo*." Those words, an innocent call for a magic door called Sésamo to open up, took on a special resonance for our group, which had spent the last ten years trying to open the country, ten years of doors being slammed in our faces and on our fingers, doors being battered down by the secret police, every door in Chile either be-ing closed or pried open, until we had managed to create a crack through which we had all managed to somehow return. That's what had done it, first the whispers and then defiant speeches, and now shouts, demanding to be rid of the forty thieves, and this was what each of us was thinking, because our one hundred jab-bing fingers were emphasizing the *ábrete, ábrete, Sésamo*, we were screaming our disguised political slogan and our undisguised emotions, gathering energy for the doors still left to unlock, we were celebrating our invincibility now that we were together. Be-cause this was not only the first time those who had remained in Chile were able to dance with exiles like us; it was also the first time they had been able to dance with one another at all in a pub-lic place.

The wild young women who had enticed us to this carnival took the lead again, with a different game: at one corner of the dance floor, the friends of the bride-to-be grabbed her and threw their victim into the air, barely catching her on the way down.

Our gang was not going to be outdone.

Viera-Gallo was soon flying up, up into the air. I was laughing my head off at his gasping protests, laughing even though I knew that it would be my turn next — and then I looked up and saw why he was pleading to be set down. Just above us, a fan hung from the ceiling, turning ominously a couple of inches away from the flying body. I watched in fascination as the tossers counted out eight, and nine, and . . . ten, and Viera-Gallo was finally safe, limbs intact, and, oh no, it was my turn next!

I tried to warn them, the fan, the fan, be careful, hey, look, look up there. But nobody paid the slightest attention. Up into the air with me!

I see one of my feet flail, approach the fan, then recede as I begin my descent. I point to the fan, gag and sputter, and I am on my way up once more, and this time the fan blades seem even closer. Roars of exultation answer my nervous hoots, what a welcome, eh, Ariel? I can visualize the tabloid headlines: RETURNED EXILE KILLED BY FAN. I'd escaped death at the presidential palace in 1973, I'd survived all the iniquities and mortifications of exile; at least let me die in some epic, solemn way, not like this. Now they've reached seven and that lucky number pacifies me. I attain a plateau of acceptance just as both my feet land squarely on the ground and it's time to go home, but the evening is not quite over.

Back in our frosted-glass-doored room, my pal Manuel Jofré hands me a book of my poems. "A time to live and a time to die," he says with a shake of his red hair, "and a time for poetry."

"Listen, I'm in no condition to — "

"We've bribed the orchestra" — and, in effect, a moment later another hush settled in. "I think they're on our side, just working-class stiffs, and they're giving us three minutes, no matter what the owner says — the Pinochetista bastard."

I began reading "Testamento" to my friends, my most notorious poem: "When they tell you I'm not a prisoner, / don't believe them. / They'll have to admit it / someday. / When they tell you they released me, / don't believe them. / They'll have to admit / it's a lie / someday, *algún día.*"

As I got to those words, *algún día,* the orchestra started up again, playing a slow, low bolero, a way of bowing to the pressure of the owner while at the same time showing solidarity with us by choosing something mellifluous and relatively quiet. *Pasarán más de mil años, muchos más,* the crooner sang softly, a thousand years will pass, more, many more than a thousand years.

Suddenly it all made sense, a pattern seemed to be emerging. We had belted out *ábrete, ábrete, Sésamo,* demanded that the doors be opened and the thieves be judged, and here was the answer, a thousand years would have to pass before . . . what? Before there was justice? Before the doors would open? Before we'd see each other again? Before there could be true love? As the volume of the music implacably rose, I began to scream my verses: "When they tell you / I'm in France / don't believe them. / Don't believe them when they show you / my false I.D. / don't believe them. / Don't believe them when they show you / the photo of my body, / don't believe them."

It was as if I were calling from exile, from one thousand years away, determined to be heard. That sonofabitch owner of the restaurant wanted to silence us? Well, my voice would carry over the wafting melody of the bolero, reaching every last patron of La Querencia. "Don't believe them when they tell you / the moon is the moon, / if they tell you the moon is the moon, / that this is my voice on tape, / that this is my signature on a confession, / if they say a tree is a tree / don't believe them, / don't believe / anything they tell you / anything they swear to / anything they show you, / don't believe them."

Reaching out to an audience I had never expected, to the husbands and wives celebrating a silver anniversary, the girls saying farewell to their betrothed buddy, the birthday people, and the revelers simply out for a good time—in the other vast room they were not dancing, only listening, the way you listen to a train by the tremble of the tracks before it emerges from the distance, they were listening and only the music was louder and louder, whipped up by the owner, I could picture him almost like a satanic conductor in a frenzy, and he wouldn't like my ending, these words: "And finally / when / that day / comes / when they ask you / to identify the body / and you see me / and a voice says / we killed him / the poor bastard died / he's dead, / when they tell you / that I am / completely absolutely definitely / dead / don't believe them, / don't believe them, / don't believe them."

no les creas

no les creas

no les creas.

The applause came not only from those in our room, but also from the other side of the doors. While we'd been ostracized and persecuted, those people, unknown to us, had spent these ten infernal years living and partly living, spectators or less than spectators, dancing to trivial tunes, blowing out candles on a cake, giving thanks for small graces. They had been there all the time, beyond our speeches and even our plans to convert them. And now their celebrations and ours had spilled over, invaded one another's spaces, destroyed the immobility reigning over their lives. Everyone there, each of the men and women welcoming us, each of the exiles who had just returned, we had been self-contained ghettos, all of us, monasteries in the midst of barbarians, we had somehow survived, stayed alive long enough to reunite now, on this dance floor, begin to cross-fertilize, sing again.

In all the years before that event, and all the years since, I've never quite experienced the mix of styles, dreams, desires, classes, as at La Querencia that night. It was a glorious anticipation of what the country could become if we managed to rid ourselves of the dictator — and of our own inner dictators, the censors inside, if we managed to open ourselves to what wasn't planned, dared to create a community as rich and plural as the one formed over the last minutes on the dance floor, if we could let ourselves go and risk the fan shredding us to pieces as we kicked down every last door in the universe.

This is what I knew, that is what I swore never to forget, that it is worth trying for that taste of heaven, no matter what the cost, no matter how dangerous it may be, no matter the noise trying to drown you out.

That is what spoiled me and cradled me and nursed me, that vision of a home for us all.

That's what really hurt, then, on my return to Chile in 1990, continues to hurt each time I go back to a country that seems farther and farther from that communal dream of solidarity. That's what grieved me on the trip in 2006, a sense of mourning that simply will not recede. The certainty that it wasn't me who had been betrayed by Chile, but rather that Chile had betrayed itself, its better self, the wondrous Chile I could not forget, I will not forget, not in a thousand years.

Fragment from the Diary of My Return to Chile in 1990

NOVEMBER 10

I wish I'd been able to tell my parents what's worrying me, sit down and out and out tell them, but they just left for Argentina after a brief one-week visit and I never broached the subject.

They love coming back to the Chile where they spent so many happy years, where they still have a large group of friends — and they've discovered a delightful little hotel in Providencia, the Hotel Orly. Though Angélica has invited them to stay with us, as they invariably did during our years of exile, my mother — no matter how much she would like to be close to her grandson Joaquín every minute of the day — understands that the house in Zapiola is really not ready to receive guests like my parents, who need many comforts in their old age. Angélica has too much on her hands as it is, Ariel, my mother said to me, you should be concerned about her, this return has really exhausted her.

For a while, as we wandered through Europe and then the United States, my parents toyed with the idea that, when we came back to Chile, they might return here as well, but the truth is that they have settled into their Buenos Aires *querido* and can't contemplate another exile.

Besides, my dad has so much to do back there. He's eighty-three years old and still going strong. After the Argentine dictatorship fell, he became one of President Alfonsín's economic advisers, though he refused to cooperate with Carlos Menem, Alfonsín's successor, wanted nothing to do with neoliberal policies imposed by Washington, preferred to create, with a group of younger economists and engineers, a think tank that could provide a critical perspective on the Argentine economy.

It's been great having them here, to bathe for a few days in their affection, revel in the fact that we finally inhabit, if not the same city or country, at least the same continent and latitude, only two hours away by plane. Just in case . . .

For as long as I can remember, there it was, the dread of their death. When I was young, I'd conjure up the day when I would sit by my mother's sweet side and hold her hand as she said goodbye, the night when my father went, quietly or raging, into the dark, and always in that picture I would be next to them, Angélica is there, our children, the whole family, for the passing, in death as in life, always, always close. Exile disturbed this vision, suggested that they might die in an empty room, whispered that I would hear of their ultimate illness when I was far away, that I would not be there to keep each of them company as they crossed whatever river of sorrow and oblivion that awaited them, I prayed that my migrations would cease before that moment came.

And now here we are, back in Santiago, just as we had prophesied, the family together, as it should be, as it has been for the long history of humanity, since the first burial mounds millions of years ago when our ancestors learned how to deal with the impossible uncertainty of death, that mystery explored and affirmed in so many multiple ceremonies, that sacred duty towards the past and the future, that connecting moment, the last hello before the first real goodbye. Now here we are in Santiago, and we . . . and I . . .

I did not tell my Chebochy, my Chúa, that we have been thinking of leaving Chile for good. I did not tell them because I dare not admit it to myself. Even to write this here is to make it intolerably real, plausible, something that may indeed happen. Talking is different, Angélica and I have been talking about this possible new migration, in some sense it has been corroding us since we arrived, as soon as Chile proved to be such a grueling country in far too many ways, we've been tossing around alternatives. What would it mean, what would be the consequences, how would I deal emotionally with a new departure after dreaming of this return for such a long time? What would it mean for Angélica to once again be separated from her mother and her three siblings? We have worried the subject, but no resolution has been reached. I have been too busy with the play, and the play is rooting me to Chile, giving me hope that it will perform some sort of miracle, burrow me into the ground of the country as if I were a wayward tree, help me avoid a new exile.

Several times I was about to tell my parents that we might stay in the United States once we leave Santiago in January, it was on the tip of my tongue, knowing that they would encourage us, support our decisions as they always have, always selflessly willing to sacrifice their happiness for ours.

But I couldn't confide in them, not this time. How can I tell them that there is a lonely dawn in the future that awaits them and me, how can I tell the man and the woman who gave me life that I may not be there when they most need me, that I will not be there to say goodbye?

•

SMACK IN the middle of our 2006 trip to Chile, we were startled by the news that General Pinochet had suffered a serious heart attack.

My first reaction was incredulity. Just as my poem advocated, I didn't *believe* him, didn't believe anything he said, or anything any of his cronies said about him. He had already faked dementia in order to escape the extradition sought by Spain when he was arrested in London in 1998. He had fooled the British authorities who were quite willing to see their unwanted guest depart even if, as soon as Pinochet arrived at the Santiago airport, he stood up from his wheelchair with both hands in the air like a triumphant pugilist.

So I was more than justified in doubting this new "sickness" of Pinochet's, especially as it happened to coincide with yet another major indictment in Chilean courts — more than a hundred had been filed and accepted by a judiciary shamed into acting by the British arrest. But he was ninety-one years old and his physicians seemed genuinely concerned, so Peter Raymont and I decided to head for the Hospital Militar de Santiago.

I had two encounters there with Pinochet supporters, both of them women. The first was at the back of the medical facility. As I was talking to a group of journalists, a slim, elegantly dressed woman passed by on clacking high heels and spat out an insult, calling me *Comunista asqueroso*, a dirty, revolting Communist, didn't answer when I asked her receding back, not aggressively, what I had done to merit such an attack. She just swished away around the corner to the front of the building, where other Pinochet zealots were crying out for their dying leader, led by a small, chubby woman, lips smeared with lipstick, fingers clutching a portrait of her hero, a litany of tears streaming from behind incongruous dark glasses. Here was a woman making a pathetic spectacle of herself for all the world to see, defending a man who had been indicted by courts abroad and in Santiago as a torturer and a thief, accusing her adversaries of being ungrateful, of having *mala memoria,* the wrong sort of memory.

And yet I was inexplicably, uncontrollably moved by her misery. I waited until there were no reporters hovering over her lamentations, so absorbed in my feelings that I forgot I was being filmed by our own crew as I approached the woman and told her, in a soft and courteous voice, how I had mourned Allende and therefore understood that it was now her turn to mourn her leader—but also wanted her to realize how much pain there was on our side. If we could acknowledge the errors of Allende, could she not do so with the terrors of Pinochet?

This sequence of the film, where I speak to that woman as she listens with surprisingly quiet attentiveness, is the one that, particularly in Latin America, calls forth the most criticism. How, people ask, could you do that? How could you validate that woman's grief for Pinochet, honor it as similar to your grief for Allende? How could I extend my sympathy to an enemy who had celebrated our sufferings, caused by her dying hero? What possessed me? That's what people keep asking.

The right word. In effect, I found myself *possessed.* It was as if some deep turmoil or angel inside had overwhelmed me.

Psychologists have discovered that a baby will cry more forcefully and for a longer time when she hears the distressed cries of other children than if the baby hears a recording of her own sobbing voice. Think about it: a baby is more upset by the voice of someone else's agony than by her own troubles. The baby intensifies her cries in solidarity with the other, shares the pain, signals to the other child that he is not alone. For me, this is proof, if we ever required it, that compassion is ingrained in our species, coded in the circuits of our brain.

This is how we managed to become human, by creating the conditions for a social network where the anguish of others is intolerable, where we need to comfort the afflicted. It is certainly not the only thing that defines us as humans; we are also characterized by cruelty and selfishness, apathy and avarice. But each of us can decide what defines our primordial humanity, and I have come to conclude, come to choose, after a lifetime of confrontations, that our most important trait is the preeminence of empathy with others. That's what constituted the base for our evolution, what lay the groundwork for our search for language, whose very core is the articulation and belief that someone else will accompany us through life. Compassion is at the origin of our species' quest for the imagination with which we can smuggle ourselves into and under alien skin. What possessed me, then, was quite simple: I felt sorry for that woman.

Even so, that act of mine needs to be interrogated. That hysterical woman, after all, was ranting against those who have *mala memoria*, targeting people like me, this very memoir, the act of remembering Allende, refusing to forget the crimes of the General. It is her memory against ours, and there is nothing I can do in this world — or doubtless in the next one — to change what she recalls, what she has selected to recall in order to defend the identity she has built for herself. Her narrative, her most intimate story, the myth by which she has lived for decades, is that Allende was a socialist who threatened her peace and property, and thus Allende's barbarian followers had to be violently suppressed by substitute father Pinochet.

It's hard to open a dialogue with people like that, as proven by the harridan at the back of the hospital who insulted me and then hissed away, gave me no chance even to approach, as I did when I saw a crack in the barrier erected by her fellow Pinochetista of the lipstick-smeared lips. Of course, in return for telling her that I understood her distress, I asked that she also try to put herself in my shoes, realize that I was not affected by a *mala memoria* but merely memories that did not coincide with hers, and that this was not, however, a reason to kill or detest each other.

It was an encounter that in some way culminated a search for answers about coexistence that had started when I had sat close to Jaime Guzmán at the concert and had fled from his presence, left him to enjoy the balm of Beethoven by himself, a search that continued after the coup and deep into exile. It could almost be ventured that my reaction to that woman was a response to my own self, the enraged Ariel who had assaulted that fascist librarian Hübner on the steps of the Hotel Des Indes in The Hague, and the disgusted Ariel who had jogged in Chile and did not know who was an enemy and who was a friend, and the puzzled Ariel who had asked in Washington how you cohabit with those who have conspired to destroy your dreams. My encounter with the wailing Pinochetista was a culmination because, from *Death and the Maiden* onward, I had been meditating extensively, in plays and novels, in poems and essays, on the walls that separate us from those who have done us irreparable harm, I had compelled my characters to deal with their nastiest antagonists and ask themselves how to avoid the sweet trap of victimhood and retribution, I had suggested in my most recent play, *Purgatorio*, that atonement was essential for any significant dialogue to transpire.

But when it came to real life, I could not wait eternally for that repentance. In real life, I felt the urge, if only for a minute, to break down those walls myself, leap across the divide, imagine a different sort of world. It's possible that my heart may have misguided me, but if that inconsolable woman was blind to reality and I had one eye open, a glimmer of an eye open, wasn't it crucial that I take the first faltering step, in hopes of unearthing some sliver of kindness that must be hidden inside every enemy's heart. Or would it be better to jail my opponents, kill them, exile them, start the violence all over again? Isn't it necessary, then, for each of us to disarm ourselves, conquer the fear of our nakedness, admit that none among us is so perfect or saintly that we are immune from the temptations of dominance?

Which doesn't mean I was offering reconciliation or forgiveness to that Pinochetista fanatic. For a long-term ceasefire to exist, some remorse would have had to bite inside, she would have had to be willing to inhabit my memories, to accept what someone like Carlos the carpenter had been through during twenty-five years of trying to keep alive his own river of memory in the midst of the conflagration. I would want her to recognize his right to show his portrait of Allende publicly, as she does her portrait of Pinochet, without fear, in front of the cameras of the world. I would want her to know what Adelaida had to go through, recognize that it's a tragedy for our country that a daughter should mourn her father in this way and know him only through photos and the memories of others, that she should feel compelled to transmit her sorrow to the faraway grandchild. I would want that fascist mourner of Pinochet to acknowledge Adelaida's right to justice, our right to remember, our right to mourn.

That woman was undeniably very far from that state of grace. But we did create, she and I, some minimal space for a minimal understanding, a gentle interlude — and history has shown us that those truces when ardent foes begin to speak to each other can be the start of something miraculous. You do not arrive at such armistices effortlessly, you often need to drive your opponents to the table through force and cunning, you cannot suppose that such meetings of the mind will simply happen. Each small step is fraught with peril and false enticements and perverse illusions. In fact, on the Monday after I had met up with that sobbing woman and reached some sort of understanding with her, I had a different sort of altercation with Patricio Melero, an ultra-right-wing congressman from Jaime Guzmán's Opus Dei–inspired political party.

We had both been invited to appear on a television program to respond to the question of whether General Pinochet, still agonizing inside the military hospital, should be afforded the honor of a state funeral. When I said that he didn't even deserve a military funeral, because he hadn't been a true warrior, had never given his enemies a chance to bury their dead, as Achilles had in that founding epic of Western civilization *The Iliad,* Melero called me vindictive, unwilling to get over the past. I fired a question back: When did you first know that people were being tortured in Chile?

He floundered, first claiming that he'd never known about torture during Pinochet's government, had only recently heard about those "excesses," and then went on to flagrantly declare the opposite, that he and his political party had intervened over and over to stop people from being tortured. When the moderator tried to move into less uncomfortable territory—like so many Chileans who prefer to look the other way, he wanted to bury the past so he could go on with his life and his program—I asked my final question: Who served the coffee?

Because *el General* did not act alone.

Beyond the many who pulled the trigger or plunged the knife or attached the metal clasp are those who bought the materials with which those horrors were perpetrated, those who kept the accounts and balanced the budget for the purchases, those who rented the basements and cleaned them out, those who paid the agents' salaries and typed out the reports and confessions, and indeed, served the coffee and cookies when the heroic combatants wearied of their marathon task, and yes, those who averted their eyes from the pain.

For those less visible accomplices, responsibility is more easily denied, making it even harder to engender those instances when adversaries meet and reach at least some sort of agreement, a pact to resort to dialogue rather than carnage to impose their points of view. Far too frequently those moments close just as abruptly as they open and we find ourselves yanked back to where we began. Even if that wall of denial is splintered for the snap and crack of a minute by someone like me, there will be no further progress unless the other side, people like that woman who insulted me, like the woman who was closeted in her anguish over the impending death of a tyrant, there will be no significant change unless people like them, like that congressman who knew his compatriots were torturing and did nothing to stop it, manage to take a step of their own, realize that to admit their collusion in the crimes is a way of liberating themselves from their own prison of prejudice and hatred.

How can this be done? Or maybe: Can this be done at all?

On a trip to South Africa in 1997 I was given a glimpse of an answer, of how brief truces between enemies can be made to last longer than a minute, become part of a country's major reckoning with itself.

South Africa is a country that, like Chile, like so many other nations around the world, has suffered severe repression and then experimented with a negotiated transition to democracy and a Truth and Reconciliation Commission, a country that therefore offered me, on that visit thirteen years ago, a chance to ask, in a different context and history, some of the same questions that had been worrying me over the years.

Eager to see how you commemorate the past so it does not die, I asked to be taken to the District Six Museum in Cape Town, that site of conscience erected in a multiethnic neighborhood torn apart by discrimination, its inhabitants uprooted and rendered homeless, inner exiles scattered to the winds of their own homeland. As I toured the museum with one of its guardians, he told me about a recent hearing of the Truth and Reconciliation Commission. A policeman of Afrikaner origin admitted killing the parents of a child and expressed regret for his actions. When the grandmother of the boy asked him what would happen when she was dead, who would care for this orphan, the policeman answered, after a pause, "Then I guess I will have to take the child home with me."

It is a wondrous story that has stayed with me.

That policeman was embodying a model of behavior, was informing the grandmother and the eavesdropping world beyond her that if we cannot undo the damage of the past, we can at least strive to undo the damage to the future, to prove by our actions today and tomorrow that we've learned from the terrors and sins of yesteryear. What other way to pay for the lives of a mother and a father than to bring home the child you orphaned, what other way to pay for a life taken than to give a life back?

And as a metaphor and epic drama, what more could we ask for, what better challenge to a world ripped apart by ethnic strife and war? I can think of no better plea for a multiracial, omnilinguistic home. Is that policeman not speaking across continents and time to the woman who years later cried for Pinochet? Is he not demanding that she take Adelaida and Paulina and Carlos the carpenter home with her, make them — and my pain as well — part of her life? Are we not all being invited to bring into our homes what is concealed behind the walls of our identity, that which most disturbs us, those memories from the thickets of others that we have considered to be alien, hazardous to our integrity? Is it not in that back-and-forth process of offering a refuge to those who are different that we can find intimations of what it means to reconcile or at least glimpse a pale path towards tolerance?

To find this path to peace is not a quest confined to Chile or South Africa or other lands that have suffered tyranny in our times. As we hurtle towards extinction, threatening to take with us our brothers and sisters and fellow creatures on this planet, it has become a task for all of humanity.

Because we are all in this together.

Maybe that is why I am writing this memoir, why it may matter to draw some lessons from a life so full of wanderings and conflicts. So I can send out this plea, teach this incredibly simple conclusion: we must trust one another. Despite all the loss, all the betrayals, we must trust one another or we shall all, all of us, surely die.

Fragment from the Diary of My Return to Chile in 1990

DECEMBER 20

It has been close to impossible to put the play on in Chile.

At first it seemed as if we would run into no problems once Julio Jung and María Elena Duvauchelle (coincidentally, two of the actors saved by Chris Reeve when they were threatened by death squads back in 1987) decided to stage it.

They were very enthusiastic, but had a question about the ending. It is what took longest to write, that last scene, where I interrupt the moment when Paulina is about to kill — or not kill — the man who may — or may not — be her torturer, the moment when she asks herself and Chile, What do we lose by not killing one of them, what do we lose? and then I have a mirror come down and we are in a concert hall and the audience is looking at its reflection while Gerardo explains how the commission's report, finally delivered, has given a voice to the dead, and then Paulina appears in evening dress and quietly sits down. Paulina, who has found her own voice during the play, is now silent, perhaps content with the small victory of listening to Schubert after years when his compositions would make her vomit, triumphant inasmuch as she has had her say, even if it was in the confines of her living room.

At this point Dr. Miranda enters and sits in the same row, all three of them now posed in front of us, looking straight into the audience as if in a concert hall, all of us, characters and spectators, part of one dreadful reality, reflected in the mirror of each other and in the mirror behind Paulina and Gerardo and the doctor. And then the *Death and the Maiden* quartet starts to play, and Paulina and the doctor glance at each other across the chasm that separates and joins them, them and us, they lock eyes — it is not clear if her tormentor is alive or dead, there in the flesh or as a ghost haunting her and us — they lock eyes and then look straight into the audience and into the future for an intolerably long time as the music plays and plays and plays.

María Elena and Julio understood that I wanted to break the illusion that what we've just seen is merely fictional, so the spectators can wonder which of those characters represents them, if Paulina or Gerardo or the doctor, or maybe the spectators are all three of those fragments of one country. María Elena and Julio agreed that this is a way of accusing us all of trying to cover up the tragedy we are living, the uncertainty we are trying to escape since the dictatorship ended.

Still, they were doubtful. A mirror coming down? From where? How to get this device without making the play prohibitively expensive — this is Chile, Ariel, and not Los Angeles, where hundreds of thousands of dollars are going to be exhausted to put on *Widows* next year. But their main concern was whether this resolution, worked on diligently with Rodrigo's help, would not leave the audience so dissatisfied and frustrated that the play might be a commercial failure, Julio and María Elena suggested that this could be one provocation too many.

I was adamant. If the spectators don't like the ending, then they should stream out of the theater and change the world, give that story a different ending in their own lives!

My passion was inflamed by what seems to me to be the underlying intellectual failure of our transition: a fear of complexity. I don't condemn it entirely. The past was so confrontational and divisive that people are, quite reasonably, wary of inciting more discord. But I abhor that gray unanimity and, above all, its hypocrisy — yesterday on television I saw congressmen from the left frivolously dancing with their right-wing colleagues in Parliament. It was for some noble cause, crippled kids, blind kids, autistic kids, certainly better to mambo alongside each other than to take out a gun and start shooting, but it still riles me, perhaps because they boogie in order to hide their substantial differences, everything swept under the rug. I abhor this sort of "reconciliation," because it plays to applause, feeds into the indifference, the lack of difference, the fear of difference. Indifference: a way of shielding oneself from the pain, of not letting the evil you have witnessed touch you. Pinochet has never had one ailment, bikes several miles every morning, lifts weights, boasts that he has *una salud de hierro,* is strong as steel. That's the essence of evil: to take the grief inflicted and transform it into muscle, vigor, potency. Whereas many former political prisoners, relatives of the Disappeared, exiles, and torture victims have cancer, terminal illnesses.

So it can break you down to dwell in pain forever, and the question is how to give yourself up to that pain, invite it into your life and give it a path out of that life without letting it destroy you, without savoring its sweet destructiveness, becoming addicted. That's Paulina's struggle, to leave that agony behind and yet not lapse into the narcotic of forgetfulness. My play might provide some preliminary answer to Chile's search for what to do with the grief, that final scene a way of establishing a zone of liberty, an ethical and aesthetic space that does not compromise, as the politicians must, in order to keep the soldiers in their barracks, keep the past from upsetting the future. But not me. I didn't come back to Chile in order to negotiate away Paulina's story.

Defiant words, but each day that passes makes it look less likely that *La Muerte y la Doncella* will ever get staged, at least in Chile.

The first crisis snaked up the day we had our inaugural reading of the play, in early November. Something was wrong:

there was so much tension between Julio and María Elena that I thought she would end up really trussing him up, I felt he wanted to strangle her, and then Julio suddenly stood up and left, said he couldn't take this anymore, and it turned out that he and María Elena were splitting up, seeking a divorce after twenty years of marriage.

And that was that. We had to replace Julio, and then replace his replacement, and then our second male actor also opted out. Meanwhile, I flew to London, to the ICA, for the reading of the play, and Harold Pinter and his wife, Antonia Fraser, were there, and Peggy Ashcroft, and Peter Gabriel and Albie Sachs from South Africa and my beloved John Berger — and I got to watch the play with a first-class cast. So I'm convinced, after that London performance and the audience's passionate reaction, that the play can travel, has real legs.

Making me more determined, when I came back to Chile a few weeks ago, not to accept any more cuts that our committed director, Ana Reeves, was demanding in order not to overly offend the audience. We were rehearsing well with the two new substitute male actors yesterday in the modest theater María Elena rented on Vicuña Mackenna, and it seemed auspicious that it should be located almost directly opposite the Argentine embassy where, seventeen years ago, I sought asylum, one more circle in my life that was closing.

And then, as happened so many other times during his regime, General Pinochet decided to prove that the only circles that close in this country do so with his consent, that he can barge into any circle and any theater and bark out his orders and we have to obey or face the possibility that *la muerte* will not be merely words in a play, *la muerte* will stalk us all over again. He was angry that a congressional committee was about to investigate the money stolen by his son-in-law, which could lead to revelations about the millions Pinochet himself was said to have squirreled away. So he placed his troops on high alert, and there was a rumor that a new September 11 coup was in the offing, and both our male actors went pale and told us they were no longer willing to be part of a transgressive experiment where we'd be putting Pinochet on trial, at least symbolically, both of them capitulated to the fear that continues to corrode Chile.

So now the play will not be on before I leave, in three weeks, for my semester at Duke, we have to find two new actors, we have to see if the lease for the theater can be renewed, we have to scrounge for new financing. My dear friend John Friedman back in the States has offered to put some money into the production, bless his soul, but I said no, if we can't get Chile to support us, then why stage the play here? But the truth is that none of my contacts in the democratic government are answering phone calls, several activists at organizations prominent in the struggle against Pinochet have refused us funds to take the play to the shantytowns or on a tour of the country.

I wonder if all these obstacles do not constitute a warning that maybe it is not the time, after all, to force my compatriots to look in the mirror and see what we have become, what we still might avoid becoming if we can all burn with the truth that I have written. What if I am wrong to try to stage the play now, when the wounds are so fresh?

•

EXACTLY ONE WEEK after my encounter with that sobbing woman outside the Hospital Militar, General Augusto Pinochet's heart stopped beating. The doctors gave all sorts of medical reasons, but I was sure those reports only partly explained his demise, did not account for the astounding coincidence that, of all the days on which he might have died, fate decreed that he do so on December 10, 2006.

December 10: the day each year when the world celebrates the adoption by the United Nations of the Universal Declaration of Human Rights. So I preferred to ignore talk of strokes and clogged arteries in a ninety-one-year-old body and proclaimed instead that the *Desaparecidos* finally hunted down their nemesis, finally deprived him of the freedom to choose the date of his own death as an answer to his refusing them life and liberty for so long.

I like to think now, four years later, that is what happened, perhaps because he did, after all, escape my ultimate dream of justice for him. Since I don't believe in the death penalty, I had devised another sort of punishment, that those eyes of his would be forced to look into the black and clear eyes of the women whose sons and husbands, fathers and brothers, he kidnapped and disappeared, one woman and then another woman and then one more, each one telling him how their lives were maimed and ravaged by an order he gave or an order he didn't block, I wanted the fierce circle of his crimes penetrated so he could understand what he had done and perhaps, who knows, one day ask for forgiveness.

Though this redemptive plan was not what came to mind as soon as I heard the news that Sunday afternoon. Nothing, in fact, came to mind. I was shocked into speechlessness, could do nothing but stare for three blank hours at the television set. My consternation was possible because Peter Raymont and his crew had left Chile the night before, were not there to demand that I comment on this extraordinary event. Rodrigo, however, about to leave that evening for the United States, captured me with his own camera. You can see me stunned, dazed, so much so that I declined my son's invitation to accompany him downtown where thousands of families were gathering to celebrate.

Rodrigo brought back exceptional footage, and he showed me some of it before he headed for the airport. I tend to be wary of any attempt to turn anyone's death, no matter how despicable the person, into an occasion for joy, so I hoped that what was being welcomed was not one man's death but rather the birth of a new nation. Dancing under the mountains of Santiago, the multitude chanted over and over, "*La sombra de Pinochet se fue*," the same phrase echoed and repeated. His shadow is gone, we have come out from under Pinochet's shadow. As if a thousand plagues had been washed from this land, never again the helicopter in the night, never again the air polluted by sorrow and violence. For those, most of them young, who were celebrating, it was as if something had been definitely, gloriously shattered when Augusto Pinochet's bleak and unrepentant heart ceased beating. They had spent their lives, as I had spent mine, awaiting this moment, this day when the darkness receded.

Maybe she was right, that future mother, seven months pregnant, who jumped for joy in the center of Santiago and the center of Rodrigo's video—was she right when she shouted to the seven winds that from now on everything would be different, that her child would be born in a Chile from which Pinochet had forever vanished? Or, I wondered as I watched her on the screen, had the battle for the soul of our divided country just begun?

After all, how often have I written lines like that one, hoping against hope that he had not eternally contaminated every schizophrenic mirror of our lives?

I was tired, that's the truth, of saying goodbye to Pinochet.

I'd been doing so since my 1983 visit to Chile. On the evening before we were to depart for the United States, my then brother-in-law Nacho Aguero had taken Rodrigo and me to a *población* to meet a young Chilean street fighter who was resisting Pinochet. On the way back, our car was brought to a halt by a screeching siren and a hive of braking motorcycles. Habitually calm and composed, Nacho this time seemed unhinged with excitement.

"*Es Pinochet, es Pinochet,*" he murmured.

A caravan of black cars raced by, and then, as they passed, a white-gloved hand darted out of one of the windows and waved. It was one more instance of Chilean surrealism, a dignitary acknowledging cheers from a nonexistent throng.

And then Pinochet was gone. An apparition.

My antagonist, of course, had no idea that I was there, watching. Yet I could not shake the sensation that the General was mocking me, that his ghostly hand in the dusk was gesturing defiantly: I am here to stay, this is as near as you and your kind will ever get to me, this is the only farewell you will ever see from me, I am as far from justice as I am from your hungry eyes. As the motorcycles vanished and the siren faded into the gathering darkness, I could hear his voice in my ear as if he were really speaking to me directly, personally, right there, nearby.

It was a voice I carried into exile with me. Haunted by it ever since that afternoon in August of 1973 when I answered the telephone at the presidential palace of La Moneda and heard, on the other end of the line, the rasping growl of *"El General Augusto Pinochet Ugarte,"* as he impatiently identified himself. He wanted to speak to Fernando Flores, Allende's chief of staff, the man who had brought me to work for the president, to see if I could help devise a media and cultural strategy to stop the coup, though the one who would really thwart the military takeover was the commander in chief of the army, there on the other side of the telephone wire. I rapidly passed him on to Flores — deaf to what that voice of Pinochet was hiding, the betrayal he was devising, the coup that had already happened in his mind.

That was the extent of our contact: no more than a fleeting telephone conversation. Maybe that was why, for all the satanic dimensions I attributed to his voice in my exile, he had remained strangely ethereal, almost disembodied. I could not dispel the sense that he was hiding, that he had been hiding perhaps from his own self all his life, learning from an early age not to tell anybody who he really was, the person he could someday become. Slipping away. Only corporeal for me in those three brief innocent seconds when I had registered his voice over the telephone, again remotely real ten years later, in the three brief vicious seconds it had taken him to wave that hand of his, that phantasmagoric glove, everything in my life and at the same time, nothing.

As Nacho accelerated away, I looked back, as if half expecting Pinochet to return, to explain himself, to materialize once and for all. But no, I was left only with his mockery and his shadow, as I was to be on the December Human Rights Day in 2006, when he was gone and I was left with all those fractured visions of him — the disembodied voice, the glove shrouded in white — still hoping that they were intimations of a possible farewell.

At the time of his death I thought, I could not help but think, Is it really goodbye this time?

Now that four years have gone by, I have come to the conclusion that the answer depends on the future, on how Pinochet's stature will be interpreted many years from now, on who will tell that story of his — which is the ultimate test of where anybody's journey ends up.

In a sense, our dictator's legacy depends on the history of the world.

If we believe that democracy and development are incompatible, if we believe that an underdeveloped country cannot attain modernity and progress unless it is ruled with an iron fist, if we believe that it does not matter if a bit of blood is spilled and certain human rights violations are inevitable in order to guarantee peace and security, then Pinochet will persist as both bogeyman and paragon to be imitated far and wide. The phrase that invariably crops up when people in beleaguered lands live in fear and turmoil, "What this country needs is a Pinochet," will be repeated, will endure.

On the other hand . . .

During my years in exile, unable to restrain what Pinochet himself was doing to me and my loved ones, I was fascinated by the possibility that perhaps we could, in some way, determine at least how the word *Pinochet* would be transmitted to generations to come.

So let me, now that he is dead, hazard a prophecy: of all the battles of his interminable life, the one that the General can no longer win is the battle for the way he will be remembered, how the hard syllables that form his name will become solidified in tomorrow's vocabulary.

We have to trust in the memory of our species, it's that simple.

I think of the children of the future, thousands of years hence, I can see them playing in a meadow or a playground.

And then one of them does or says something that warrants a reproach, an insult, a hideous slur, from the other one, who shouts out, "Oh, don't be a Pinochet."

"Pinochet?" answers the other. "Pinochet? Who's that?"

Pinochet?

Who in hell is Pinochet?

So was that the last circle that needed to be closed, in my life and in the documentary about my life? What better ending, after all, than the General's demise? No, there was another circle, more resonant with me and with the world than this petty despot's departure from the earth, one more bit of unfinished business that needed attention, one more experience and place to film, this time in the United States rather than Chile.

We were headed for Ground Zero.

That "hallowed ground," which had added a second tragedy of terror to my existence, could not be missing if we were to tell the whole story. We could not omit the country where I had been bred as a child, the country from which I had been exiled at the age of twelve, the country that had contributed to my exile in 1973, the country where I now write these words, ten blocks from where my two American granddaughters breathe their joy into the world and into our lives.

So we set out, with cameras at the ready, for New York, another city of my dreams assaulted on another September 11, again a

Tuesday morning when fire fell from the sky. Though by 2001 very few people in the world recalled the existence of that remote Chilean date, I was besieged by the need to extract some hidden meaning behind the juxtaposition and coincidence of those twinned episodes bequeathed to me by the malignant gods of random history. There was something horribly familiar in that experience of disaster, confirmed during my visit to the ruins where the twin towers had once reached for the sky.

What I recognized was a parallel suffering, a disorientation that echoed what we had lived through in Chile. Its most turbulent incarnation was the hundreds of relatives roaming the streets of New York after 9/11, clutching photographs of sons, fathers, lovers, daughters, husbands, begging for information, are they alive, are they dead?, every citizen of the United States forced to look into the chasm of what it means to be *Desaparecido,* with no certainty or funeral possible for those who are missing. The photographs were still there in 2006, pinned on the wires separating the ogling spectators from the abyss, encouraging me to use the unique perspective of my own life to forge a message to the citizens of America lost in a labyrinth of pain.

Call it a gift from Chile to the nation that did so much to destroy our democracy, the nation that was also mine, the America where I thrive and teach and write, where my Isabella and my Catalina will grow.

We Americans — yes, we — received that day all of a sudden the curse and blessing of being able to look at ourselves in a way habitually denied to most of our citizens, the chance to distressingly imagine ourselves as part of the rest of humanity. Never before had they — yes, they — been ripped apart to this degree by the ravages of guilt and rage, the difficulties of memory and forgiveness, the uses and abuses of power, the true meaning of freedom and responsibility. And consequently never were Americans more tempted to apply amnesia to their yesterdays and innocence to their tomorrows, never was it more perilous and easier to sweetly, vindictively rid themselves of the complexity and contradictions of their newly naked predicament.

Chile, for all its imperfections and failures, found a way of responding to the terror inflicted on us (yes, us, we Chileans), a path of peace rather than war, a path of understanding rather than retribution. A model that the United States, wrestling with the mirage of its imperial ambitions, did not have the immediate wisdom to follow. And yet the complacent invulnerability of this nation where I now abide has been fractured forever, as the gash in that site at Ground Zero reveals. We citizens will have to share, whether we wish to or not, the precariousness and uncertainty that is the daily lot of the majority of this planet's other inhabitants. A crisis of this magnitude is one of those opportunities for regeneration and self-knowledge that are granted, from time to time, to certain nations. It can lead to renewal or destruction, used for aggression or for reconciliation, for vengeance or for justice, for the militarization of a society or its humanization.

One of the ways for Americans to go beyond the insecurity that has been swallowing us since 9/11 is to admit that our suffering is neither unique nor exclusive. If we are willing to look at ourselves in the vast mirror of our common humanity, we may find ourselves connected with many apparently faraway men and women who have trekked through similar situations of injury and fury.

A message to America I was able to deliver with more forcefulness in the documentary because I had recently become a U.S. citizen.

I had resisted taking that step with as much passion as I had put into trying to remain in Chile during our unfortunate six months in 1990. The implacably practical Angélica had decided to seek naturalization soon after we resettled for good in the States, and then hauled our two sons to Charlotte, North Carolina, for the interviews and swearing-in ceremonies.

I was a tougher nut to crack. I had already switched allegiances twice before — from Argentina to the States and from the

States to Chile — and damned if I was going to relapse a third time, especially now that physical absence might weaken my ties to Latin America. Though my obstinacy had more intricate reasons.

No matter how much I might proclaim my mission to be a bridge between the Americas, the voice I had created for myself, the persona I projected, was that of a Latino from the South. I derived authority, power, credentials from that outsider status, relished being a sort of unofficial spokesperson for those who could not make themselves heard from our derelict lands. I had grown comfortable with that tone and viewpoint. It served me well on television and radio, in my op-ed columns and interviews, in readings at bookstores and commencement speeches, a deepening of the perspective I had discovered that morning in Bethesda watching the snow that was and was not mine fall silently. It had crusted into a second skin, become a home away from home, struck the right balance by allowing me to intervene in both Chile and the States from a middle point of intersection and detachment. And each time after 9/11 that I faintly contemplated reconsidering Angélica's arguments in favor of nationalization, something would flare up, in Santiago or Mexico or some neglected corner of Latin America and the words would come flying, in English and in Spanish, and I didn't want to squander that — there is nothing more difficult to abandon than a voice.

And then had come the arrest of General Pinochet in London in 1998, and his year and a half of captivity, and all of a sudden my public persona was more valuable than ever, on the BBC and *Charlie Rose* and Chilean TV. You see, I said to my wife, *ya ves,* if I were an American citizen, how could I possibly write publicly to Pinochet and tell him that this was the best thing that could have happened to him, that he has been afforded an implausible chance to repent. It is only feasible to write words like those as a Chilean, that's why I could write to an unknown Iraqi dissident in the *Washington Post* and say that I understood why he wanted to be rid of the tyrant Saddam but not at the price of an intervention from abroad, explain that I would have rejected such a solution for my Chile in the days of our dictatorship, even if it had meant that friends were to die. I felt that my role as a public intellectual depended on keeping my distance from any official association with a United States misruled by George W. Bush, that Chile was more relevant than ever, the glass darkly through which I saw torture and the erosion of civil rights and "extraordinary rendition," again the outrageous familiarity. I had grown accustomed to the idea that the United States, with all its blemishes and shortcomings, was a haven against persecution, at least for someone like me, and now it was threatening to turn into a police state, foreigners were being rounded up, permanent residency was no guarantee against abuses and Guantánamo, my Lord, and Dick Cheney, no longer a congressman receiving my copy of *Widows* in 1983, was churning out real widows all across the oil homelands of the planet twenty years later.

Angélica would hand me clippings, as if I couldn't read, as if I didn't know: Listen to this provision of the Patriot Act, no, Ariel, I want you to listen. And also: You want to be effective? Then break out of the snug cocoon, say *we* when you speak to Americans, include yourself in that *we*.

And she was worrying, my wife, *Escúchame,* Ariel, if they expel you, I'm not leaving, this time I'm not following you, you want to never see your granddaughters again? Angélica would not give up. It was absurd, there was no chance of anything of the sort happening to me, not with my contacts, not with my profile, not with — it can't happen here? Wait, wait, hadn't I written, just last year, that it can happen anywhere, make people afraid enough and they'll let the government do anything in their name?

So the day came when, for Angélica's birthday in 2005, my present was a card inscribed with a promise: *Mi amor:* I am ready to become a citizen. *Feliz cumpleaños.*

And my succinct explanation to all who wondered how the

revolutionary Ariel, the voice of the oppressed of the Third World, could commit the sacrilege of embracing the America that had cast me into exile and was torturing prisoners in Abu Ghraib — Pinter asked me, but why, Ariel, why would you do that? — I had a good answer: I would rather go to jail again than into exile one more time, I will never again go into exile.

It's even partly true — or at least the excesses of the Bush years made my exculpation sound reasonable at the time. Just as when I'm asked, Why don't you live in Chile?, I have a pithy response ready: I left my country because I could stand that they didn't like me, but not that they hated my work. So much more complicated, the scattered reasons for my many expatriations, but what am I supposed to do, expound at length what this book has barely been able to stammer, what those six months in 1990 and the seventeen years of incessant migration revealed to me? We all use shortcuts and templates to tell our life story, are constantly refashioning it so our yesterdays accommodate what we are doing today, how we remember ourselves tomorrow. Not a bad sound bite: I decided to become a citizen so I wouldn't be treated the way Bush's America was treating the rest of the world.

And nothing changed, by the way, Angélica had been, as usual, right. At times I still gag when I say we, it does not come to me readily, it is easier to leave a country than to forsake a thought deeply embedded in the neurons of a brain — but this change in status has been an educational experience. I've grown accustomed to this life on the hyphen, a Chilean-American born in Argentina tightroping between loyalties, as divided and united as my libraries. Nothing much has actually changed except that I leave Miami with my U.S. passport and pull out my Chilean passport when I enter Pudahuel airport, now no longer stopped by anyone with an order to deport me though still irritated by the prosperous countrymen so blatantly indifferent to anything but their own pleasure. And then, with my Argentine passport, I travel to Buenos Aires, less, alas, than I did when my dad and my mom were alive, but I often go to visit my sister Eleonora and friends and editors and the boulevards of the city where I was born and where I am always received with more sympathy than in Santiago. Everything will be fine for this wanderer who is at the intersection of all those overlapping communities, as long as there is not a war between these nations laying claim to my body.

And there was something more than convenience and pragmatism when I stood in a courtroom in North Carolina on a summer day in 2006 among all those other men and women from every unlikely corner of the world and took the oath of allegiance to this nation conceived in liberty, that man who didn't believe in oaths or allegiances or nations. Not bogus what I felt, that tremor. Real emotion. Inside me was the boy who had watched the Manhattan skyline slip below the horizon as we headed for Chile, saying goodbye to the Statue of Liberty, which mocked him but also beckoned across oceans and time and nostalgia, *Come back to me,* not only English had been waiting but America as well.

In the days before I handed Angélica her birthday card, I had been coming to terms with my ambivalence, I was composing a love letter to America, a Whitman-like prayer in the form of an imprecation: Let me tell you, America, of the hopes I had for you. Hopes that had not been met, but still a love letter, asking America to beware of the plague of victimhood, the surge of self-righteousness that comes from being unfairly hurt, beware the plague of fear and rage that had consumed me for so many years.

Those words of love poured forth, above all, I suppose, from a renewal of my adolescent faith in that other America. The America of *as I would not be a slave, so would I not be a master,* the America of *this land is our land, this land was made for you and me,* the America of all men and all women, every one of us on this ravaged, glorious earth of ours, all of us, created equal. Created equal: one baby in Afghanistan or Iraq as sacred as one baby in Minneapolis. Where was my America? The America that taught me tolerance of every race and every religion, that filled me with pioneer energy? Was I wrong to believe that there was still enough rebelliousness and generosity in America, enough citizens unspoiled by excessive wealth and false innocence and imperial overreach, to conquer its fear? Wrong to believe that the country that gave the world jazz and Eleanor Roosevelt, Cassavetes and Toni Morrison and Philip Marlowe, would be able to look at itself in the cracked ice of history and join the rest of humanity, not as a separate City on a Hill, but as one more city in the valleys of sorrow and uncertainty and hope where we all dwell?

And that faith in America was vindicated, partly vindicated at least, two years after I became a citizen.

I had danced on the streets of Santiago on November 4, 1970, when Allende had become president, and then danced once more on November 4, 2008. Thirty-eight years to the day after my fellow citizens in Chile celebrated the inauguration as president of a man who believed in social justice and peace, my fellow citizens in the United States voted along with me for a man who would stop his country from torturing in the name of security, chose as their president a man who bore on his skin the marks of a race that had been oppressed as our Chilean poor had been oppressed from time immemorial.

So again I danced in the streets, this time the streets of this mongrel nation about to be governed by a mongrel like me. I danced even if inside me the specter of Chile was murmuring to be wary of too much enthusiasm, that change in the United States would face, as it does everywhere, the rabid opposition of those who will not give up their privileges without a violent struggle. I knew, of course, even as my body bobbed and weaved and jumped up and down, I knew that the country of Barack Obama would have to tackle many of the dilemmas we lived through in the country of Allende and Pinochet, in our own unsuccessful revolution, and during our own constricted transition. If you go too fast, like Allende, you risk a disastrous end, I whispered to the president-elect, almost as if he could hear me, you may not change the world after all, but if, like the inevitable Enrique Correas of our time, you advance too cautiously, friend Obama, you risk losing your soul, you may not make much of a real difference, you will lose the enthusiasm and inspiring vision necessary to fight for any deep and lasting modification. I had seen the Chilean revolution founder in the seventies because of its inability to create enough support among the people, I had seen a regime of avarice and terror defeated in the eighties, only to watch how in the nineties it kept a twisted grip on the multiple levers of power, I remembered in every one of the cells of my dancing body how we had been outmaneuvered by the forces of the past, our desire for true democracy cornered and poisoned by fear.

But that did not take away the wonder of each victory, small and large, nor was saying goodbye to the America of Bush and Reagan and saying hello to the America of Obama and Roosevelt and Lincoln the only reason why I danced. I was dancing also because that was my body's way of saying and swaying that this land could also be my home, one of my many homes, my joy was not only for the world and for the United States but for this citizen who was accepting his life, with all its turmoil, with all its misfortunes and displacements. I was accepting my life as it had been danced.

Final Fragment from the Diary of My Return to Chile

JANUARY 6, 1991

Today I am leaving Chile.

This evening I will board a plane and say goodbye to this land I have called my own for most of my adult life.

Over the past six months I have refused to admit that this day would arrive, a day that replicates the journey I took in 1973, as I head north again, but this time without imagining a permanent return to the south. Except in my head, except in my literature, except in the skip of my heart as I look back. Oh, we'll keep this house, I won't put these books away in boxes, not yet, maybe not ever, somewhere inside me the fantasy stirs that this departure may not be definite — what if they love my play, what if it manages to shake Chile up and prove that the country is open to my offerings? But no, this decision needs to be recognized for what it is: a capitulation. I waited for a miracle, fought hour after hour to hold on to my identity and my country, I fought and lost.

I'm succumbing to the curse that tracks the men of the family, even if Angélica sees it differently, feels that by returning to the States we'll end all this wandering, that we should not sacrifice Joaquín's happiness for *este país de mierda.* Thus far I have stubbornly responded that it might be a *país de mierda,* but it's my *mierda,* my damn shit, a glorious *mierda* that I was destined to drown in or drain and purge. Until gradually, over the last couple of weeks, as the day of departure for the States to start teaching at Duke drew near, my decision has grown easier to bear.

Basically, I forced myself to acknowledge this truth: I don't like who I have become here, what Chile is turning me into, someone feasting on incessant irritation. I can't stand the trickle of daily vitriol being deposited inside me, this is not how I want to live the rest of my life.

This unexpectedly became clear to me when I attended a funeral on December 28, this fateful 1990 ending with yet one more death.

Ana María Sanhueza had been so alive when I had visited her a few weeks ago! That hepatic cancer devouring her body had made my dear classmate from university days even more beautiful, molded her usual radiance into that of a yellowing, magnificent Asian princess. I was grateful for this chance for us to spend time together, after all those births I had missed,

first moments and last moments, all those baptisms and fare-
wells. As we chatted quite normally for several hours, thanks
to her astonishing lack of fear, I realized how seldom I had
thought of her during these seventeen years of absence, and
yet, guardian of my memories, she had summoned me to her
bedside so that she could take one last one, one last memory
with her to the other side of death, and she was kind enough
to restore some to me for safekeeping. There was truth, then,
to the myth of Penelope, weaving and unweaving her cloth
every night, maybe there will always be someone back home
who is waiting for us at our own simulacrum of Ithaca. Ana
María contained a tapestry of shared experiences that I had
forgotten, we laughed so hard and enjoyed each other so much
that I finally had to be evicted, promising to come back before
the year's end. And then the news came that she had died on
Christmas Day.

Saddened as I was, I thanked her for this one last gift, that I
could say goodbye to her dead body just a few weeks after having said goodbye to her living smile, and at the Cementerio
General I was able to reconnect with former classmates, some
of whom I hadn't seen in decades. Among them was Raquel
Olea, a feminist literary critic who had married and then di-
vorced one of our colleagues. I knew she had not been dealt
too many lucky cards in life, so when she asked how things
had been going, I simply opened up, didn't hide a thing, perhaps because we had never been that close I confessed how
rough this return had been. Five years, she said, echoing the
exact number Antonio Skármeta mentioned to me, it will take
five years before you'll be accepted here. And meanwhile, you
have to use your elbows, Raquel said, open space for yourself
with sharp elbows, Ariel.

I didn't tell her I was planning to leave, that the halfhearted
decision had already been made, but answered instead that
I didn't have five years and didn't want to use my elbows, I

haven't got elbows left, sharp or soft or long or short, that's
what I vehemently poured out to her as the year ended.

Elbows. Such an unromantic, overlooked piece of our anat-
omy, and standing in that *camposanto* where I had imagined
my own last rites would someday take place — how often had
I beseeched the gods of exile to save me from the fate of dying far from home — standing there I looked at Raquel's elbows and looked at Ana María's coffin about to be turned into
dust, and the reluctance that had been holding me back was
completely swept away. I was suddenly as weary as Angélica
and Joaquín and Rodrigo, and as convinced, with my whole
heart and every ventricle and valve and river of blood running
through my arms and into my elbows, that I could no longer employ the *empanadas* and the mountains and the anonymous *pueblo* that thrilled me and the parking attendant who
called me his *amigo,* I couldn't use any of that to counter the
unhappiness of my family, my own unhappiness, nothing was
enough to keep me here, not *La Muerte y la Doncella* and not
all the hopes I had projected onto the land and not my Spanish vernacular in all its colloquial glory and not my parents
nearby in Argentina, it was just not enough, not lunch with
Queno or dinner with Pepe or an excursion to the Cajón del
Maipo with Antonio or a visit from Elizabeth Lira, not a sweet
conversation with Isabel Letelier or a chess match with Santiago Larraín, just not enough, Angélica's incredible family and
the future children whose birthing I would miss, it just wasn't
enough, I couldn't grow the elbows I needed in order to open
that space to breathe. No, it wasn't enough that Allende was
buried just a few lanes away, so close to where I had wanted
to be carried and taken and tumbled along these tombs, not
enough that I would give up resting with the dead who had
kept me alive in exile, it just wasn't enough anymore, the many
tiny triumphs of our *pueblo,* just not enough, I didn't have one

millimeter of an elbow left, I didn't want to jab anybody in the ribs, that's not how it was supposed to be.

It's time to get the fuck out of this country.

So is that it? Has the Chile I fell in love with really died?

Yes and no, I say, and also: It's not your fault, *no es tu culpa.* I'm the one, I expected too much from you, *mi país,* too many ways in which I wanted you to save me yet again from the de-mons of my loneliness.

I chose to disregard the warning signals over the years.

On the last afternoon of our first tumultuous return to Chile in 1983, I had stopped by my brother-in-law's house to say goodbye and found myself alone for a few minutes with Patricio's wife, Marisa. She had asked me, So when are you re-turning? And my answer was tentative, as always, when the question had come up: as soon as we could.

All of a sudden, Marisa began to cry. Taking one of my hands, tears pouring down her face, she said, "*No vuelvas, huevón.*" Don't come back, you idiot. And added, more in-tensely, "*No vuelvas nunca a este país de mierda.*" Don't ever come back to this fucking country.

Barely fifteen minutes had elapsed since she had been de-vising all sorts of schemes for our return. The most recent of other flare-ups. In the middle of something else, our closest friends would inexplicably start berating the country as if they hated it. Not Pinochet, not the military, not Nixon, not the corporations. Chile itself. Declaring that if they could leave, they'd pack their bags tomorrow. Stay in Bethesda, Ariel, An-gelita. And then, a short time later, as if they were amnesiac, casually suggesting the safest sort of houses to live in when we got back.

These moments were a crack in the smooth glass of our welcome, fleeting but alarming enough to allow us a glimpse of what my loved ones had been through, what they were trying to hide from me or from themselves, a reminder that our family of four would be returning not only to the Resistance but also to the dread and the betrayals, the lassitude and the shallowness, the confusion of past mistakes and the repetition of old vices.

It is a crack that has widened on every return and now has ruthlessly installed itself inside me, the realization that deep within some of the very angels and comrades in Chile who have resisted the worst were the same slouching fiends I had brought back inside myself from exile, the perdition I had hoped to leave behind when I returned to my fabled land. If I fear that monsters have corroded the soul of the country, it is because exile has created in me monsters of my own, or revealed that they had always been there, *agazapados* inside the sad, crouching truth of what I had done in order to survive, just like the country, just like everybody in the country. And now I will have to carry those imperfections back with me when I leave Chile, I will have to concede that Chile did not have all the solutions.

Concede that I may be able to better find that spiritual evolution far from here. If I were to come back, give it one more try, it would again be day after day of a dig here and a poke there, day after elbowing day, I'd never be at peace in a place like Chile unless I ignored the overwhelming tide of what exasperates me, unless I turned into somebody who was either silent or overbearing, exasperated precisely because everything here was so mine, so recognizable.

No te metas. Don't get involved. Take a step back. That's what Angélica has kept advising me during our many returns here when Pinochet ruled, she would frequently rein me in, remind me what one mistake can cost, how fragile is the thread of life. But she couldn't ask me to hold back entirely. During those years in exile, I had prayed for Pinochet to fall and prayed as well that it would not happen before I had a chance

to plunge into the whirlwind of the Resistance, discover first-
hand the experience received only by hearsay, derivatively,
from reports and articles that always arrived late, and from
phone calls with bad news in the middle of the night that always arrived much too early, I begged that I would not be forever chained to the memories and audacity of someone else.

So how could I not *meterme*, intervene, when I saw something unfair, an outrage, how to become a bystander? One
incident, just one: One morning in the center of town I had
come upon an eight-year-old girl performing in front of a
crowd. She dipped a rag in paraffin, lit it with a match, then
snuffed out the fire by cramming the rag in her mouth, repeated the operation again and again, ceasing only when a
nearby man signaled her to collect the few pesos dropped in
a hat on the ground. As for me, I lost my cool, began to be-
rate the throng, how could they just stand there and watch and
not lift up a voice in protest, *hasta cuándo*, till when, till when,
what does it take to move you, make you care?

They dribbled off, none of them responding, not even the
pimp and his waif, everyone making believe I hadn't said a
word. *No te metas.* Don't get involved, what Pinochet hammered into us all day long, each man for himself, the same
suggestion replicated by the democratic government for sun-
dry, unrelated reasons. But at least in dictatorial times fear
could justify inaction. We now have no such pretext, unless we
grant that apathy has made us all into accomplices.

So . . . *Así que me he metido.* I plunged in. To the hilt. Under the dictatorship a host of issues not patently political, social issues that galvanized me, had been postponed because
of their possible divisiveness, but democracy was supposed to
be different, we should know better, we had fought precisely
for a society that was more just and equal and tolerant. That's
why, upon this 1990 return to Chile, I propelled myself into
a permanent campaign for all sorts of rights, almost a cru-

sade to change how people feel and think and act in their everyday lives. With abortion outlawed, no divorce laws, prim and proper manners stiffly enforced, Chile continues to be a mummified, hierarchical land, maids mistreated and children shunted aside and the handicapped hidden from view and indigenous people derided. Pinochet's seventeen years have inflamed the habitual Catholic fear of sex, encouraged and increased our chronic homophobia — and not only among supporters of the dictatorship, so every social gathering is speckled with moments of potential discord. "Just don't take the bait," I tell myself afterwards or beforehand, don't ask the maid to sit at the table, don't offer to clear the plates, don't take every joke against *maricones* as a personal affront, don't remind the hosts who have just said there are no *indios* in Chile that there are at least a million of them, that they are the real custodians of the earth, don't comment on how my compatriots degrade the environment and leave trash everywhere, *no te metas,* Ariel, don't be so prickly, don't say that you want to greet the autistic adolescent isolated in the back of the house, don't, don't, don't, me and my big mouth.

I've often asked myself if one of the reasons I fought so hard to stay behind in Chile after the coup was because I saw it as an opportunity to change my demonstrative personality that gets me into so much trouble. In our house in Durham, Angélica has pasted next to my desk a phrase from the Prisse Papyrus, the oldest book in the world, words attributed to the vizier Ptahhotep of Memphis in the year 2350 B.C.: "Let your ideas fly freely but keep your lips sealed." If I had remained in Chile after the coup, I would have been forced to obey that Egyptian overlord from the fifth dynasty of the Middle Kingdom, become like the men and women who stood up to the enemy with implausible valor but paradoxically bit their tongue in everyday life. My compatriots have learned to bide their time, to develop a subtle spider web of jokes, slight allegories and innuendoes, a language that has survived the dictator, that continues to molder our daily routine. They have spent a near lifetime eluding the direct confrontations that I feel are essential for a true democracy to flourish and that marked my life in exile. That's what I thought would make me valuable here, that I'd find a way to reconcile what I'd learned during my banishment with what citizens in Chile had learned while under the boot.

As if the two halves of my life could so easily be harmonized.

Just as I carried into exile the voices of liberation that had sung the Allende revolution, it was inevitable, I guess, that I regenerate in Chile itself the voices and lives accumulated while I wandered. I can no more repudiate them than betray the voices of the *Desaparecidos*.

It would take more time than I have now, as the hour of my departure draws closer, to spell out how I have changed, the many new visions swirling inside as I escape the snare of Chile, but one example should suffice: Deena Metzger.

As close to a sister as I have in this world.

Deena had come to Allende's revolutionary Chile from California with her then lover, David Kunzle, seeking out the author of the Donald Duck book that he wanted to translate into English, and we had sat in a café. When it started to rain, I remarked that the rain made me sad, because the independence day festivities were coming up and our people needed to dance, and Deena stood up and danced in the rain, and when the coup came and interrupted our dance of multitudes, she implored the gods to care for us, for each of the *compañeros* she had met during her visit here. And every one of them survived.

The day after the coup, Deena sent me a telegram from Los Angeles, cannily reminding me that I had an invitation (that she had invented out of thin air, of course) from a California university, when should she send tickets for me and the family? Throughout the years of exile, Deena has educated me about the earth and the feminine soul of the universe and our lasting connection to animals and how to greet the rain in the midst of the rubble, how to begin again.

The United States I will be returning to is not only the land of Nixon and Reagan and Bush. It is also the land of Deena the healer — and I don't need to add to her name all the other names and bodies, the mesh of voices in English and in Spanish that will give us a homecoming, smooth our return to wonderful Duke and the land that I know as well, perhaps even better, than this country I am now abandoning.

Maybe that's why I even entertained the possibility of going back permanently to the States once the idea began to surface, maybe that was what gave Joaquín hope that our decision to remain here was not as final as we had proclaimed, maybe it was the attraction of what that outside world had to offer.

Let's set aside the day-by-day chronicle of these six months, not debate whether the streets needed to be repaired or were secretly resisting modernization, whether Chile was best represented by the unresponsiveness of the government or by the dignity of the workers expelled from their jobs, not hold up the exaltations of my library in Santiago against the sorrows of Joaquín and the frustration of Angélica and the departure of Rodrigo. That's how I have written this return up to now, as a choice between two Chiles, one pulling me towards it and the other pushing me away, exclusion or inclusion. But underneath that tug of war a different confrontation was taking place, two versions of my persona were contending for supremacy.

Like every exile in history, I was consumed with one desire when I left Chile in 1973: to return home intact, unaltered, indomitable. All too soon, also like every other exile who has wandered this earth, I discovered that in order to survive in a foreign land I would need to bow my head and adjust to the world outside. That impediment to my plan for purity was intensified in my case by the circumstance that I was living my third major deracination. After having lost Argentina and then the United States, I was now facing severance from the land I had chosen for my adult destiny. The fact that I was already multicultural and bilingual when I set out on my journey, and did not have, as did Angélica or the great majority of my fellow refugees, childhood roots in Chile, increased the temptation to accommodate to an outside orbit. That is why, I think, I stayed away for so many years from the lure of the United States, ended up there only by accident. Because I was afraid that once I was in America, I would never go back home.

It was in order to contest that fear that I forged, when I departed from Chile after the coup, a narrative about myself, a story that came to dominate the way I understood my role as an intellectual and a revolutionary.

It was a tale emanating from the depths of Chilean history. In the same year, 1541, that Santiago was founded by the Spanish conquistador Pedro de Valdivia, an Araucanian youth called Lautaro was captured and, like so many other males of his tribe, made to serve. The boy must have been exceptionally bright and handsome, because Valdivia took him on as his personal page. Lautaro stayed with that new master for seven years, long enough to appropriate the techniques and language of the foreigners, before escaping back to his people. Five years later, he led an insurrection, which ended with Pedro de Valdivia ambushed and killed. It was the knowledge of those foreign weapons and vocabulary that gave the Mapuches their victory. Indeed, for more than three centuries, until Western industry and science broke the technological balance in Chile with the Remington rifle (as it did all over the globe, from India and Africa to the American West), that race of warriors was able to fight the invaders to a standstill.

Born at the beginning of the modern age, when Chile itself was being fashioned out of the clash of the local and the
international, Lautaro became the first of my compatriots to
tackle the dilemma of how you live in a colonized land, and
also the first to reject the dazzling path of assimilation that so
many others took, proclaiming that resistance for the powerless is a better path to identity than becoming a servant or an
administrator or a concubine to the intruders occupying your
soil. That resistance, Lautaro said, could be bolstered and not
weakened by what those usurpers possessed. The trick was to
be cunning, return to the tribe from your forced exile loaded
with new knowledge; the trick was to remain loyal. Surging
out of the mountains of our communal past, Lautaro told me
that it was possible to be double, indeed, that to be split was
essential, given the planet we inhabit, as long as we remain
true to our genesis and to the deeper heart that is hiding back
home. This Third World, insurrectionary Odysseus told the
young man heading for alien territory that to be cosmopolitan
and to speak more than one language should be recognized as
weapons that would help, rather than hinder, his return, that it
was possible to become a rooted cosmopolitan.

That's how I narrated my own exile as it was happening,
that's the persona I defended myself with, clasped through
each of my tribulations and returns. In a world where Allende
was dead and someone like Reagan could twice be elected
president of the United States, for a writer who had skewered
Donald Duck only to discover that Disney was more globally triumphant than ever, for someone who had rejected the
dominant American model and now saw the *Yankee Go Home*
graffiti on Third World walls modified by the droll *And Take
Me with You!,* Lautaro afforded me the hope that nothing had
been substantially modified, not in me, not in my country, not
on the planet, not the long view of where history had to be
heading. So when I descended from the plane six months ago,

in July of 1990, I returned to Chile with the conviction that life back home was more worthy, meaningful, and hopeful — in a word, more real — than life anywhere else, Chile as the one place in the world where I could best defy death.

But that is not the only way to defy death, the only way to heal the world, the only story I smuggled back with me from exile. In the seventeen years that had passed, a different, parallel paradigm settled, almost unawares, almost unstated, into my life. It is not embodied in a character as dramatic as the Chilean warrior, but in a yet unnamed model of existence that I will have to explore in the decades ahead.

Looking back on my life, a pattern emerges. Each of the dislocations that determined my fate and the fate of my family have corresponded, I realize, to a tectonic shift in the battles of the twentieth century. My grandparents were part of the gigantic migratory wave in search of freedom in the decades before the First World War, all those millions deserting a continent whose nations would soon be savaging each other. And then my father had to leave Argentina because of the struggle between fascism and democracy, that crucial conflict that greeted me when I was born, smack in the middle of a second world war. And then another war, just as decisive and planetary, the Cold War, leading to yet another exile, fleeing McCarthy and intolerance and the Red Scare, off to Chile where a young doctor called Salvador Allende was already planning his first run for the presidency, anticipating other wars, world wars as well except they were fought out mostly by proxy, in the neglected and blighted backlands of our planet, in Africa and Asia and Latin America, the wars of liberation of the sixties and early seventies, our revolution in Chile being one of those, though ours was peaceful, even if it ended in blood.

And again into exile I went, again the larger forces were deciding for me, men in faraway rooms were shadowing me, moving their chess pieces oblivious to my existence, but still I had to play on their game board. And now I am about to start living and writing out the consequences of one last exile, expatriation, displacement, withdrawal, whatever you want to call it, the long hybrid aftermath that will pursue me once I depart from Chile on this Day of the Magi. I will be compelled to reluctantly embrace that final upheaval in my twentieth-century life, which was also fleshed out in world history, this last exile of mine that came to pass at a time of deeper transitions, the fall of the Berlin Wall and the defeat of Pinochet and the arrival of a system that had always been global but never this instantaneous, never this interdependent, never this pervasive, this moment that will be looked back upon in the future, if there is a future, if there is anybody to look back, this era that will be deemed the age when humanity finally became one, whether because what is made in China is bought in Kansas or because who eats in California will determine who hurts in Valparaíso. This is the time I am to become globalized along with the world, this is the time of misery and promise when fate has chosen me to journey forth from my salvation in Santiago into a world from which I will not retreat except through death.

Not alone, I will not set forth on this new journey alone. This Lautaro who now knows he will never really go back home must again carry with him the voices and images of a past that has not died. I will be free from the bracing limitations of nationality, but that freedom will not, I hope, block me from my own responsibility towards those who continue to represent the suffering of all humanity. I may be leaving Chile, but Chile, the country of Allende, will not ever leave me. My idea of it, my incessant hope for it, will not leave me alone, will keep challenging my solitude, stalk me, hold me accountable, remind me of the need to dream. Something inside me will always remember the homeless of Amsterdam and the turnstiles jumped in Paris and the lack of health care in Washington, I have to remember, as the temptation of detachment and pride intensifies, as the allure of prestige and the spotlight call again, I need to always remember the lessons of struggle and estrangement. The intellectual and emotional adventure of becoming a world citizen has to be coupled with my commitment to a literature for the pain of that world. If the best of what the world offered me has followed me here to Chile and troubled my integration into this society plagued by terror, how can I not, on the reverse journey away from the South, smuggle out something just as valuable?

The *Desaparecidos* have ghosted and saved me in exile, and I pray that their raw, savage unreality will keep me centered — or maybe it is de-centered — in the midst of the enticements of expatriation.

Soon I will be far away. I can feel the remoteness and abstraction growing inside again, once more facing the black hole of myself in a faraway country that has no simple answer to my litany of questions, this man who can no longer be consoled, this time, with the certainty that I have something to come back to.

Maybe this is as much as I can hope for, almost more than I can ask for: that I will keep inside my heart some fire with which to survive.

Heart and hearth. In English, somewhere in the collective reminiscence of the language, somewhere in its unspeakable psyche of our ancestors, the heart and the fire, home and fire, are deeply parented, like *hogar* and *hoguera* in Spanish, *foyer* and *feu* in French. When fire was extracted or nursed or stolen from the gods of lightning, it gave rise to the sedentary in humans, gave our species a center around which to converge, gave to the wheel that would eventually be invented, a place where its wanderings and turns could find repose. Outside the shadows cast by the flames, far from the meat being cooked and the pottery being shaped and the metal being forged, unseen by those whose pounding hearts shared that shelter, there were outer rims of darkness, intimations of what was unknown and predatory and called foreign because it could not be celebrated together through community and storytelling by the light of that fire. And maybe that is my final lesson, what I have come back to learn, take back to warm myself in those nights ahead—perhaps that heart of hearts that is *hogar* and home in Spanish and comes from the same root as *fuego* and fire, cannot now, maybe never can, absolve my doubts.

I may have to drag those doubts with me forever.

But also this. My exile has taught me that this heart is everywhere, that it does not belong to one country or one person or one community, it has been beating in all the friends abroad who have cared for us, literally giving us heart, their heart, when we had felt most abandoned. I believe I came back to Chile to find it again, to confirm that it had not been soiled beyond repair, was still inside the people who had struggled and inside a country to which it was worth returning.

I may have been disenchanted, but I also found that heartbeat, I am also carrying away the immense heartbeat of Paulina and her brothers and sisters inside me.

When a catastrophe that I cannot foretell strikes again in the years ahead, as it will, as it must, as it has over and over in my life and the larger life of our species, all that matters, ultimately, is that I may be found worthy of that hearth, that humanity burning in the night.

And to think that all I ever wanted was to come home.

EPILOGUE

... The final word
Is yet unspoken.

— Bertolt Brecht, "Concerning the Label Emigrant," 1937

Could it have turned out differently?

Is there an alternate vector of history that does not find me writing this epilogue in North Carolina, not writing in English but in Spanish, is there an avatar of mine in Santiago as the sun goes down and glows a goodbye on the cordillera? Once in a while I ask myself — I cannot help it — why I am not growing old in the city of my dreams.

I go back from time to time, rarely for more than a few weeks. The longest period I have spent in Chile over the last two decades was a month in 2003, when I trekked with Angélica into the foreboding desert of the Norte Grande in order to write a book. *National Geographic* had invited me to select any site in the world for the book and I chose that zone of Chile. Yes, Chile has continued to provide the primary communal prism through which I contemplate the layers of my past. And then, of course, in 2006 that documentary, when Pinochet did us the favor of dying, and many trips in between to see the family and some friends and drink in the breeze that flows in the afternoon hours of summer.

It's enough for me to step off the plane to again feel enchanted and appalled, outrageously happy and equally exasperated. And one of my reveries has come true: everything about Chile I had once hoped would beguile Joaquín now delights our granddaughters. Isabella and Catalina, who have twice vacationed with us in Santiago, run in freedom through our Zapiola community and bask in the friendliness of the people and gape at an old horse dragging a cart full of vegetables on the street, their visits cleansing the land of harsh memories, bathing it in my lost innocence.

As for Joaquín and Rodrigo, they've reached an oasis of reconciliation with Chile, and Angélica, critical as ever of the government, the glitterati, the vulgar TV programs, enjoys being in touch with the everyday folk she knew as a child. She loves her bungalow in La Reina, and to me the house is comforting, a sort of reprieve. It's worth the worry and the expense merely to know that something of ours remains fixed and immutable back there, a door still semi-ajar, at least in my mind.

Would it have made any difference if *Death and the Maiden*, when it finally opened in Chile in March 1991, had been received with even a pinch of the generosity with which it had been conceived, if the Chilean elite had not repudiated it, if Enrique Correa, like so many others in the government, had embraced the play instead of refusing to subsidize it, all those former comrades who turned their backs on me when we asked for help? Would it have mellowed my relationship with Chile if a playwright I had considered a friend had not written that my drama was the worst play written in the history of the country, would it have reconciled me to my land?

It makes no sense to ask those questions, to complain about the foul reception of a work that dragged into daylight the viral truth of Chile's crisis. I had been naïve to expect that the powerful men on either side of the political divide would adopt my impertinent literature as theirs, support a provocation that threatened to upset the delicate rules and pacts and codes of our fragile consensus. Literature can never be consensual for me. It must always demand to bring out what Pinter called "the weasel under the cocktail cabinet." It was good that I had not been *consagrado*, that they'd evicted me from the *sagrado*, the sacred, that no one had cared enough about my vision to flatter me with a deadening pantheon of praise and sinecures. Unavoidable, that clash of mine with Chile's cautious transition. Unless, that is, I had been ready to forswear the voice groomed in exile.

The one lesson I did not want to forget as I wandered: never let them take away your dignity. It was a lesson that came to me from the first recorded case of exile, the story of Sinuhé, an old bearded man, homeless and ailing, who had wandered over lands he could barely remember, until finally begging forgiveness of the pharaoh who had expelled him. Unless Sinuhé's body was oiled and bandaged and placed in a sarcophagus, if singers and dancers did not accompany him to the white marble tomb, safe passage to eternity would be blocked forever. "You shall not die on alien soil," the royal decree proclaimed. "Nor be buried by Asiatics, wrapped in sheep's skin. Take care of your body and come home."

Sinuhé's name is known only because of that abjuration, that cautionary tale of a man who, in order to die in his own land and live in the hereafter, was forced to destroy his self-respect. My foremost struggle all through my exile: not to imitate that man.

But of course immortality for me — as far as it exists — is not in the body but in art. And to have remained silently behind in Chile would have been to disown my responsibility as a writer. So there was nothing that unusual, after all, about my final exodus. Like so many authors in our troubled modern times, I had been defeated by the conservative forces ruling my society. And like them, my departure — the first time in my life I had left a land I loved because of my own choice and not because some men in the night and fog had willed it — was in the long run a form of liberation, a freeing of my life and my literature from the narrow constraints of nationality.

It is true that there are days when I still wonder if the retreat from Chile was foreordained. What would have happened if I had not been arrested at the airport that August morning of 1987 and we had been able to settle in Santiago permanently that year, been given a chance to adjust to the country before Pinochet was gone, before the transition to democracy brutally told us we were not welcome? That's all it would have taken, perhaps: for Rodrigo Rojas to walk down a different street of that *población,* for the truck with the military patrol that burned him alive to have had a flat tire, that's it . . . Is that all it really would have taken, one footstep more, one horror less, for everything to have changed? Or was the arc of my life determined at another traumatic moment, was its course already decided that day in 1945 when a child with pneumonia was interned in the Mt. Sinai Hospital ward in New York and emerged three weeks later speaking only English? If I had not been blessed with this language in which I now write these memories, I might not have held in reserve the option of leaving Chile when things got unbearable in 1990. Isn't that why I did not burn my bridges — because I was the bridge itself that would have been burned, because the man returning to the false paradise of Santiago already transported within him the seeds of a dual existence? This is what Joaquín may have noticed: the hesitation on the part of his parents, that we had not sold our house in Durham, that I had not resigned from Duke, that another existence awaited me, that I could choose to be the person I was rather than the person I had planned to be. Could it have turned out differently?

I do not repent.

Writing this book has been my therapy, *mi última palabra*, my last word on how my life turned out. These pages are my attempt to look back so I can stop looking back, finally lay to rest, in a burial made of words, a past that needs a decent funeral so something sad in me can sleep in peace.

Like my languages, also at peace, with me and with each other.

Though every reconciliation — be careful what you wish for — has its difficult moments.

Take the other night. Sitting at the dinner table with my family, I needed something, *eso*, and I opened my mouth to name it so someone could hand it to me. I waited for the word to come, but nothing, not to the tip of my tongue, not to the stem, not down the throat, and I lingered, expecting one of my two languages to take pity on me, and *nada*, nothing, still nothing, they were ganging up on me.

There I was, *pobrecito*, marooned on this island of speechlessness in the middle of the bustle of Angélica and Rodrigo and his wife, Melissa, and the little ones, Isabella and Catalina gabbing away, and Joaquín and his girlfriend Robin. My granddaughters are remembering that only a few minutes ago they were climbing Mount Abu, hilariously climbing up and down their *abuelo/* grandpa as if he were the cordillera itself, they all have a surfeit of vocabulary at the very instant when a dwarf is drilling a hole in my head, and neither the Spanish that birthed me nor the English that gave me safe haven, neither of my protector/*idiomas* hurries to fill the void even though I'm helplessly married to them both, I almost thirst to again be monolingual, simple, elementary.

If I wait any longer the food's going to get cold — and I can't stand my food cold! — so what can I do but point, my finger stretching in the direction that my tongue refuses to go, jerking at the damn thingamasomething, *eso*. I recall a theory of how language started, the sound grunting in hot pursuit of where the index finger was thrusting, and now my sons are into the game. Rodrigo picks up a napkin, "*Esto, Ariel?*" and I shake my head, and "This, Pops?" and that's Joaquín showing me a piece of bread, and they know and I know that what I want is — I'll remember later when it's in my hand — the insignificant salt shaker, *el salero*, but right now all is sudden silence, I can nearly hear the rabbits in our small forest — I always wanted to live in the middle of the woods, and there are turtles and a family of foxes once in a while and snakes and more trilling birds than anyone could identify, wordless creatures who will never find themselves gesticulating at a dinner table, they flow through life like rivers of wonder — I can almost hear the planet turning and the stars being consumed by the furnace of their own light. I am paralyzed, that word I am seeking has disappeared along with all the dictionaries, *el castellano y el inglés* have swallowed each other and reduced me to infancy, suspended me as if time had stopped and space had been abolished, turned me into the target of an Oliver Sacks essay, and if Angélica had not finally shown some mercy and passed me the object in question, pronouncing the word as if I were an idiot or a child learning to speak for the first time, who knows if I would not still be there, a baby, a pre-human mammal, a deaf-mute, waiting, waiting, waiting for the salt of the earth I could not name.

And yet isn't this something to be desired? Both my languages ripening into peaceful coexistence as they age, cohabiting so closely that they can enjoy a good laugh at my expense?

I am beset by the intuition that it was always thus, this mutual aid. It is true that I was born twice, once in Spanish as I fell from my mother's womb in Buenos Aires and once in English when I rose into the winter of New York, and true as well that I can recall neither of those meetings, the day I breathed for the first time and the day when I was so sick I almost stopped breathing. But who can doubt that words were close by, neighboring me on both occasions, greeting me as they have greeted every human on this earth. I don't know what either language said to keep me alive and pacify the dark, but I have lived long enough to trust that even back then they were working together and were joined in love for the vast vocabulary of all humanity, I have lived long enough to believe that the origin of life and the origin of language and the origin of poetry are all there, in each first breath of each inhabitant of this planet, each breath as if it were our first, each next breath, the anima, the spirit, what we inspire, what we expire, what separates us from extinction as we inhale and exhale the universe, the simple, almost primeval arithmetic of breathing in and breathing out.

The written word came later, that effort to make breath everlasting and secure, carve it into rock or ink it on paper or sign it on a screen, so that its cadence can endure beyond us, outlast our lungs, transcend our transitory body and touch someone with its waters. To those waters I have dedicated my existence, calling to those who breathe the same air to also breathe the same verses, with the certainty that we can bridge the gap between bodies and between cultures and between warring parties.

By now I have learned, I pray I have learned, that all journeys, gentle or turbulent, need to hold out the hope of forgiveness in order to really begin, in order to find the courage to end. Perhaps that is what this memoir has become, a way to forgive myself, be patient with the road taken, be thankful for what was offered and for what I was able to return to others, thankful ultimately for the multitude of my many homes as I sought to be true to that little boy who had sworn never to forget why it is crucial to imagine a world, struggle for a world, without injustice and unnecessary sorrow.

Could it have turned out differently?

Anything else would have tasted like ashes.

JANUARY 6, 2011

ACKNOWLEDGMENTS

Para qué sirven los versos si no es para la noche
en que un puñal amargo nos averigua, para ese día,
para ese crepúsculo, para ese rincón roto
donde el golpeado corazón del hombre se dispone a morir?

Sobre todo de noche,
de noche hay muchas estrellas . . .

What are verses for if not for that night
when a bitter dagger finds us, for that day,
for that moment of dusk, for that broken corner
where the shattered heart of man makes ready to die?

At night above all,
At night there are many stars . . .

— Pablo Neruda, "Oda a Federico García Lorca,"
 from *Residencia en la Tierra*, 1937

AMONG THE MULTIFOLD novels I will never write is one where a first and only chapter consisting of a brief sentence is followed by endless acknowledgments by the author that go on for pages and pages. I bring up this Nabokovian idea because if I were to minutely thank everyone who helped me write this memoir, not to mention those who helped me simply to survive so that it could be written at all, these final acknowledgments might entail a volume as long as, and perhaps longer, than the text that I have just completed.

I have tried to spare myself the temptation, and the readers the ordeal, but certain specific instances of recognition are imperative.

Angélica's role as the initial and final reader of the manuscript cannot be exaggerated. As if living by my side for almost fifty years and following me in and out of so many countries and exiles were not enough, she then had to read several times over the story of that traumatic journey, correcting details and reminiscences, and, of course, style and vocabulary.

Indeed, at night, there are many stars.

Let me also recognize the importance of those immediate collaborators without whom *Feeding on Dreams: Confessions of an Unrepentant Exile* would not exist in its present form.

Jin Auh, my loyal friend and agent who steered me through several versions and many tribulations and found me the right haven.

Deanne Urmy, my editor at Houghton Mifflin Harcourt, who believed as fiercely in the significance of this book as I did, and proved it by working selflessly to make my words clear and true. Even someone who, like me, has defined himself as a bridge, needs a bridge like Deanne for company and guidance, needs additional help to reach across the abyss of cultures and continents to our readers.

And more help, precise and encouraging and brilliant, was provided by Larry Cooper, who corrected the many defects of the manuscript with as painstaking care as I had tried to put into the writing.

Thanks also to my editor in Spanish, Josefina Alemparte, who helped me improve this text when I finally found a way to tell my story all over again in my other language.

And what can I say to Desmond Tutu, who, among so many other gifts, offered me, through an interview with my friend and collaborator Kerry Kennedy, the words with which my memoir ends: Anything else would have tasted like ashes.

Indeed, at night, there are many stars.

As for everybody else, this coda will resolutely refrain from mentioning anyone who has already been named in the book itself. If you have appeared in any way, shape, or form, please take this as a sign of deep gratitude. And those many friends whose company I would have liked to include in this memoir, whose stories are part of my story and who simply did not fit in the body of the text, I wish to show a minor token of my appreciation by listing you all below, in alphabetical order and without any undue embellishment as to why and where and how. Their fleeting presence here, at the very end of these dreams I fed on, is my way of telling them that they have not been forgotten.

Indeed, at night, there are many stars.

Justo Alarcón/Jorge Albertoni/Josephine Alexander/Mariano Aguirre/Rafael Alberti/Isabel Allende Bussi/Robert Alpaugh/Pedro Altares/Valeria Ambrosio and Gonzalo Falabella/Srinivas Aravamudan/Yunit Armengol/Anna Ashton/Margaret Atwood/Pat Aufderheide/Sun Axelson/Noemí Baeza/Rubén Barreiro Saguier/Pía Barros/Lluis Bassetts/Drina Beovic and Arturo Infante/Bob Berger/Chuck Bergquist/John Biaggi/Soledad Bianchi and Guillermo Núñez/Betsy Bickel/Jim Billington/Sergio Bitar and Kenny Hirmas/Augusto Boal/Tracy Bohan/Cecilia Boisier/Linda Brandon/Phil and Betsy Brenner/André and Karina Brink/Richard and Cindy Brodhead/Moisés Brodsky/Ian Brown/Tito Bustamante/Michael Byrne/Peter Carey/Uwe Carstensen/Bernard Cassen/Julio and Zulema Castellanos/Ascanio Cavallo/John Cavanagh/Patricia Cepeda and John O'Leary/Bill Chafe/Amalia Chaigneux/Martine Chaltiel/Mischka Cheyko/Marybeth Chiti/Claus Clausen/Glenn Close/André Codrescu/John Coetzee/Dianna Cohen/Furio Colombo/Marcelo Contreras/Maria Cordon/Félix Córdova Moyano/Antonio and Chichina Cornejo Polar/José Antonio Cousiño/Mark Cousins/Sebastian Cox/Achmat Dangor/Ruda Dauphin/Gordon Davidson/Ann Dekker/Raquel de la Concha/Poli Délano/Alberto Díaz/Ryan Dilley/John Dillon and Johanna Melamud/Leonard and Rhoda Dreyfus/Max du Prez/Bob Egan/Rodrigo and Manena Egaña/Atom Egoyan/Joe Eldridge and María Otero/Mark Ellam/Eve Enseler/David Esbjornson/Joaquín Estefanía/Jaime Esteves and Jacqueline Weinstein/Patricia Fagen/ Mariano Fernández and Queca Morales/Roberto Fernández Retamar/Hamilton Fish/Peter Florence/Rogelio Flores/Juan Forn/ Ambrosio and Silvia Fornet/Jonas Forssell/Cathy Friedman/ Verónica Fruns/María Pía Fuentealba/Bill and Mary Fulkerson/ Lizi Gelber/Gary and Pela Gereffi/Graeme Gibson/Gaby Gleichmann/John Glusman/Nick Goldberg/Pablo González Casanova/ Lou Goodman/Sue Goodwin/Nadine Gordimer/Ariane Grasset/Philip Griffiths/Claudio Grossman/Edith Grossman/Jorge Guzmán/Peter Hagan/Peter Hakim/Jaime Hales/Patricio Hales/ Peter Hall/Clark Hanson/Regina Harrison/Jorge and Norma Heine/Nick Hern/Christopher Hitchens/Dominique Hollier/Diana Holtzberg/Michèle Hozer/Gerardo Ilabaca/Luke Ingram/ Rebecca Irvin/Deborah Jakubs and Jim Roberts/Deborah Karl/ Danny and Lynn James/Judy James/Fred Jameson/Tom Keneally/Kerry Kennedy/Ranji Khanna/Ted Koppel/Barbara Koppel/Peter Kornbluh and Eliana Loveluck/Claudio Kuczer/Margaret and Phil Lawless/Clare Lawrence/Tina Ellen Lee/Martha Lefevre/Daniel Leiva/William Leogrande/Francisco Letelier/Denise Levertov/Pei-Fen Liu/Jens Lohmann/Ao Loo/Bridget Love/ Abe Lowenthal/Kevin Lynch/Mercedes Lynn de Uriarte/Amin Maalouf/Humberto Magistris/M. Mark/Sergio Marras/Judy Martínez/Tomás Eloy Martínez and Susana Rotker/Gerald Marzorati/ Kay Matschullat/Mike Medavoy/Hugo Medina/Régine Mellac/ Gil and Karen Merkx/Cara Mertes/William Merwin/Walter and Anne Mignolo/Wilson Milam/Ethelbert Miller/Santiago Moledo/ Víctor Monasterio/Chloë Moncomble/Ben Mordecai/Marcio and Marie Moreira Alves/Alberto Moreiras and Teresa Vilarós/Walter Mosley/Mario and Nicole Muchnik/Kuki Muller/Victor Navasky/ Martyn Naylor/Hugues et Simone Neel/Naín Nómez/Georgina Núñez/María Angélica Núnez/Kenazburo Oe/Carlos Olivares/

Michael Ortiz Hill/Nigel Osborne/Rafael Otano/Paik Nak-chung/
Grace Paley/Marta Paley de Francescato/Joe Papp/Bill Paterson/
Fernando Paulsen/Victor Perera/Patricia Politzer/Hedda Post/
Jennifer Prather/Nora Preperski/Marie Cécile Renaud/Carole
and Curt Richardson/Nan Richardson/Lucy Rivas/Jan Rofekamp/
Kathleen Ross and Daniel Szyld/Ricardo Sabanes/Carlos Salas/
Pedro Sánchez/Al Sandine and Mary Bradford/José Saramago/
Miguel Sayago/André Schiffrin/Alessandra Serra/Susan Sheehan/
George Shivers/Rob Sikorski/Raúl Silva and Birgitta Leander/Bar-
ney Simon/Dan Simon/Johnny and Diane Simons/Paul Slovak/
Jin-Chaek Sohn and Sung-Nyo Kim/Susan Sontag/Osvaldo So-
riano/Saúl Sosnowski/Allister and Sue Sparks/Ilan Stavans/Juliet
Stevenson/Jacqueline Stroop/Trudie Styler/Robert Thompson/
Ernesto Tieffenberg/Edward and Josefina Tyriakian/Julia Tyrrell/
Michele Uthard/Arturo Valenzuela/Augusto Varas/Alvaro Va-
rela/José Vargas/Marta Vilela/Carlos and Ana María Villagra/Fer-
nando Villagrán/Mirta Villalba/Stephen Walker/Cora and Peter
Weiss/Ted Weiss/Carole Welch/Mieke Westra/Elie Wiesel/Alex
Wilde/Susan Willis/Penelope Wilton/Pam and Bob Winton/Eric
and Hazel Woolfson/Cathy Wyler/Andrew Wylie/Eduardo Yent-
zen/Ida Zelicovich/Raúl Zurita.

Indeed, even by day there are stars, the sun is there and the
other stars.

TIMELINE

1942 — The author is born not only into this world but also into a Spanish language that eagerly welcomes him to his native Buenos Aires.

1945 — The author loses his Spanish and gains English upon arriving in New York, as he follows his family into the first of many exiles.

1954 — The author must leave his beloved United States at the age of twelve because his father is persecuted by Senator Joseph McCarthy. In Chile, Spanish and the revolution and, eventually, a charmed woman called Angélica Malinarich await him.

1960 — The author decides not to return to the States to study on a scholarship but to remain in Chile, though he will continue to write in English.

1964 — The author raucously participates in the presidential campaign of the socialist Salvador Allende. Allende loses, but the stage is set for another attempt by the same coalition six years later.

1966 — The author, having graduated from the University of Chile with a degree in literature, marries Angélica.

1967 — The author and Angélica have their first son, Rodrigo, and also celebrate Ariel's becoming a Chilean citizen.

1968–69 — The author and his family spend a year and a half as a research scholar in Berkeley, California. He reaches a drastic conclusion: to no longer write in English, the language of empire, but in Spanish, the language of insurrection.

1970 — The author dances on the streets of Santiago when Salvador Allende wins the presidency and starts a peaceful revolution that will be met with escalating violence by its enemies. The author combines his political activism with a flurry of writing, including a novel, in Spanish, which garners a major literary award.

1973 — The author manages to survive the military coup that ends Chile's democracy and Allende's life. After several weeks spent underground, he is ordered to seek asylum by the Resistance. Against his will, he goes into exile, arriving in Argentina at the end of the year.

1974 — The author and his family must flee Buenos Aires and its death squads. After visits to Lima and Havana, they end up in Paris.

1976 — The author, after two and a half miserable years in France, accepts a post at the University of Amsterdam and embarks on a creative writing splurge after a long period of silence. Before leaving Holland, he will have written four books in four different genres: essays, poems, short stories, and a novel, as well as numerous magazine articles.

1979 — The author and Angélica have their second child, Joaquín, and plan their next move, to Mexico, in order to wait out the dictator, who seems to have consolidated his hold on Chile.

1980 — The author and his family leave Amsterdam for what is presumably a year in Washington, D.C.

1981 — The author is denied a resident visa to Mexico and must stay in the United States and live off his writing as well as other odd jobs.

1982 — The author writes an op-ed for the *New York Times*, the first time he has published anything in English.

1983 — The author and his family get green cards that allow them to remain legally in the United States; a few months later, Pinochet announces that the subversive Ariel Dorfman may now come home. Two days after that, the family arrives in a Chile that is in the throes of a massive revolt. This fifteen-day visit will be followed, in the years to come, by other returns, as Ariel and Angélica explore the possibility of permanently resettling in Chile.

1986 — The author and his family move to Durham, North Carolina, after reaching an agreement with Duke University to teach one semester a year. This contract will allow Ariel to travel to Chile and buy a house there while making his living abroad for the next three years. He is in Santiago when Rodrigo Rojas, a young Chilean expatriate, is burned alive by the military. Back in the States, Ariel spearheads a campaign against Pinochet.

1987 — The author and Joaquín are arrested at the Santiago airport and deported. As a result of international pressure, Ariel is allowed back two weeks later, but a definite homecoming is postponed.

1988 — The author participates actively in the Chilean plebiscite that will decide whether General Pinochet will continue as president for life. The victory of the democratic opposition starts the slow countdown to the end of the dictatorship.

1989 — The author again returns to Chile, to vote for Patricio Aylwin for the presidency and sets in motion plans to go back permanently next year. Rodrigo, now twenty-two years old, decides to return right away.

1990 — The author goes back to Chile with Angélica and Joaquín. Although he shies away from creative fiction and public appearances, Ariel writes a journal about the return that he has been dreaming of for seventeen years. He also writes *La Muerte y la Doncella,* and the day after finishing the play, he starts on an English version that will have its first reading in London at the end of the year.

1991 — The author and his family leave Chile for good after spending six months there. *Death and the Maiden* proves to be extraordinarily successful, both as a play and as a film, directed by Roman Polanski.

1996 — The author begins to write his memoir *Heading South, Looking North*. He continues to visit Chile from time to time.

1997 — The author is on a trip to South Africa when he receives the news of his mother's death in Buenos Aires. He is too far away to attend her funeral.

1998 — The author is astounded to hear that General Pinochet has been arrested in London for crimes against humanity, and participates in the campaign to ensure that he is brought to trial. After eighteen months under house arrest, the General is flown back to Chile.

1999 — The author is in Durham for the birth of Isabella, the child of Rodrigo and his wife, Melissa. Three years later, Catalina, their second child, enters this world. The delight of Ariel and Angélica in their granddaughters cannot be exaggerated.

2001 — The author witnesses on TV the terror attacks on New York. This is his second September 11 and will determine much of his writing and activism in the years to come.

2003 — The author learns of his father's death in Argentina, which occurs on the same day that the United States invades Iraq. Ariel is able to fly to Buenos Aires in time for the funeral. Angélica was at Adolfo Dorfman's side when he died.

2006 — The author becomes an American citizen and later in the year travels to Buenos Aires, Santiago, and New York to film *A Promise to the Dead,* a documentary based on his life. During the visit to Chile, General Pinochet dies. That trip is the catalyst for *Feeding on Dreams: Confessions of an Unrepentant Exile.*

2008 — The author votes for Barack Obama for president of the United States.

2010–11 — The author finishes the manuscript of this book. It has been forty years since Allende's triumph and twenty years since Ariel definitively left a Chile where he thought he would live forever. On the day he completes the new book, he begins to rewrite it in Spanish.